CLIMATE CHA...

THE INDOOR ENVIRONMENT,

AND HEALTH

Bo ... tice

THE NATIONAL ACADEMIES PRESS 500 Fifth Street, N.W. Washington, DC 20001

NOTICE: The project that is the subject of this report was approved by the Governing Board of the National Research Council, whose members are drawn from the councils of the National Academy of Sciences, the National Academy of Engineering, and the Institute of Medicine. The members of the committee responsible for the report were chosen for their special competences and with regard for appropriate balance.

This study was supported by a contract between the National Academy of Sciences and the US Environmental Protection Agency via award No. EP-D-09-071. Any opinions, findings, conclusions, or recommendations expressed in this publication are those of the authors and do not necessarily reflect the view of the organizations or agencies that provided support for this project.

International Standard Book Number-13: 978-0-309-20941-0
International Standard Book Number-10: 0-309-20941-2

Additional copies of this report are available from the National Academies Press, 500 Fifth Street N.W., Washington, DC 20055; (800) 624-6242 or (202) 334-3313 (in the Washington metropolitan area); Internet, http://www.nap.edu.

For more information about the Institute of Medicine, visit the IOM home page at: **www.iom.edu.**

Printed in the United States of America

Cover credit: Thermal image of a residence in New Haven. © Tyrone Turner/National Geographic Society/Corbis.

The serpent has been a symbol of long life, healing, and knowledge among almost all cultures and religions since the beginning of recorded history. The serpent adopted as a logotype by the Institute of Medicine is a relief carving from ancient Greece, now held by the Staatliche Museen in Berlin.

Suggested citation: IOM (Institute of Medicine). 2011. *Climate Change, the Indoor Environment, and Health*. Washington, DC: The National Academies Press.

"Knowing is not enough; we must apply.
Willing is not enough; we must do."
—Goethe

INSTITUTE OF MEDICINE
OF THE NATIONAL ACADEMIES

Advising the Nation. Improving Health.

COMMITTEE ON THE EFFECT OF CLIMATE CHANGE ON INDOOR AIR QUALITY AND PUBLIC HEALTH

JOHN D. SPENGLER (*Chair*), Akira Yamaguchi Professor of Environmental Health and Human Habitation, Department of Environmental Health, Harvard School of Public Health, Boston, Massachusetts

JOHN L. ADGATE, Professor and Chair, Department of Environmental and Occupational Health, Colorado School of Public Health, University of Colorado, Aurora, Colorado

ANTONIO J. BUSALACCHI, JR., Director and Professor, Earth System Science Interdisciplinary Center, University of Maryland, College Park, Maryland

GINGER L. CHEW, Epidemiologist, Division of Emergency and Environmental Health Services, National Center for Environmental Health, Centers for Disease Control and Prevention, Atlanta, Georgia

ANDREW HAINES, Professor of Public Health and Primary Care, London School of Hygiene and Tropical Medicine, London, UK

STEVEN M. HOLLAND, Chief, Laboratory of Clinical Infectious Diseases; Chief, Immunopathogenesis Section, LCID; Tenured Investigator, Immunopathogenesis Section, National Institute of Allergy and Infectious Diseases, National Institutes of Health, Bethesda, Maryland

VIVIAN E. LOFTNESS, University Professor, School of Architecture, Carnegie Mellon University, Pittsburgh, Pennsylvania

LINDA A. MCCAULEY, Dean, Nell Hodgson Woodruff School of Nursing, Emory University, Atlanta, Georgia

WILLIAM W. NAZAROFF, Daniel Tellep Distinguished Professor, Vice-Chair for Academic Affairs, Department of Civil and Environmental Engineering, University of California, Berkeley, California

EILEEN STOREY, Surveillance Branch Chief, Division of Respiratory Disease Studies, National Institute for Occupational Safety and Health, Centers for Disease Control and Prevention, Morgantown, West Virginia

Program Staff

DAVID A. BUTLER, Senior Program Officer; Study Director

LAUREN N. SAVAGLIO, Research Associate

TIA S. CARTER, Senior Program Assistant

RACHEL S. BRIKS, Program Assistant

VICTORIA WITTIG, Christine Mirzayan Science and Technology Policy Fellow

HOPE HARE, Administrative Assistant

NORMAN GROSSBLATT, Senior Editor

ROSE MARIE MARTINEZ, Director, Board on Population Health and Public Health Practice

Reviewers

This report has been reviewed in draft form by persons chosen for their diverse perspectives and technical expertise in accordance with procedures approved by the National Research Council's Report Review Committee. The purpose of the independent review is to provide candid and critical comments that will assist the institution in making its published report as sound as possible and to ensure that the report meets institutional standards of objectivity, evidence, and responsiveness to the study charge. The review comments and draft manuscript remain confidential to protect the integrity of the deliberative process. We thank the following for their review of the report:

Patricia Butterfield, Dean and Professor, Washington State University, Spokane
Peyton Eggleston, Professor Emeritus, Pediatrics, Johns Hopkins Children's Center
Kristine M. Gebbie, Joan Hansen Grabe Dean (acting), Hunter-Bellevue School of Nursing, Hunter College, City University of New York; Professor, Flinders University School of Nursing and Midwifery
Peggy L. Jenkins, Manager, Indoor Exposure Assessment Section, Research Division, California Air Resources Board
Patrick Kinney, Associate Professor of Public Health, Division of Environmental Health Sciences, Columbia University, School of Public Health

Donald Milton, Professor and Director, Maryland Institute for Applied Environmental Health, University of Maryland
Andrew K. Persily, Leader, Indoor Air Quality and Ventilation Group, Building Environment Division, Building and Fire Research Laboratory, National Institute of Standards and Technology
Thomas J. Wilbanks, Corporate Fellow, Oak Ridge National Laboratory

Although the reviewers listed above have provided many constructive comments and suggestions, they were not asked to endorse the conclusions or recommendations, nor did they see the final draft of the report before its release. The review of the report was overseen by **Richard B. Johnston,** Associate Dean for Research Development, Professor of Pediatrics, University of Colorado Denver School of Medicine, and **Lynn R. Goldman,** Dean, The George Washington University School of Public Health and Health Services. Appointed by the National Research Council and the Institute of Medicine, they were responsible for making certain that an independent examination of the report was carried out in accordance with institutional procedures and that all review comments were carefully considered. Responsibility for the final content of the report rests with the authoring committee and the institution.

Acknowledgments

This report could not have been prepared without the guidance and expertise of numerous persons. Although it is not possible to mention by name all those who contributed to the committee's work, the committee wants to express its gratitude to a number of them for their special contributions.

Sincere thanks go to all the participants at the public meetings convened on June 7 and July 14, 2010. The intent of the workshops was to gather information regarding issues related to climate change and public health. The speakers, who are listed in Appendix A, gave generously of their time and expertise to help to inform and guide the committee's work. Many of them also provided additional information in response to the committee's myriad questions.

The committee extends special thanks to the dedicated and hard-working staff of the Institute of Medicine's Board on Population Health and Public Health Practice, who supported and facilitated its work. Board Director Rose Marie Martinez helped to ensure that this report met the highest standards of quality.

Finally, the committee members would like to thank the chair, John D. Spengler, for his outstanding work, leadership, and dedication to this project.

Contents

Summary

Climate change[1] poses "a significant long-term challenge for the United States" (NRC, 2010b). Its potential effects on public health have been addressed in major research efforts conducted under the auspices of the federal US Global Change Research Program and the National Center for Environmental Health, the congressionally mandated National Academy of Sciences' *America's Climate Choices* study initiative, and the Intergovernmental Panel on Climate Change of the United Nations Environment Programme and the World Meteorological Organization. A search of the US National Library of Medicine's *PubMed* database in late February 2011 yielded nearly 1,500 papers on the topics of climate change or global warming and health.

In all that work, one issue has been given relatively little attention: the effect of climate-change–induced alterations in the indoor environment on occupant health. At first impression, the lack of attention might seem reasonable. Buildings shelter occupants from the outdoors. A deeper examination, though, provides reasons to be concerned. People spend the vast majority of their time in indoor environments and will thus experience many of the effects of climate change indoors. The outdoor environment permeates indoors in all but maximum-containment laboratory conditions. A building that was tightly sealed as a response to adverse outdoor condi-

[1] This report uses the term *climate* to refer to prevailing outdoor environmental conditions—temperature, humidity, wind, precipitation, sea level, and other phenomena—and *climate change* to refer to modifications in those outdoor conditions that occur over an extended period of time.

tions or because of efforts to reduce energy use might protect occupants from one set of problems but would increase their exposure to another: such buildings tend to have decreased ventilation rates, higher concentrations of indoor-emitted pollutants, and more occupants reporting health problems.

Against that backdrop, the US Environmental Protection Agency (EPA) asked the Institute of Medicine (IOM) to convene an expert committee to summarize the current state of scientific understanding with respect to the effects of climate change on indoor air and public health. It provided three examples of key questions to address:

- What are the likely impacts of climate change in the United States on human exposure to chemical and biological contaminants inside buildings, and what are the likely public health consequences?
- What are the likely impacts of climate change on moisture and dampness conditions in buildings, and what are the likely public health consequences?
- What are the priority issues for action?

This report, prepared by the Committee on the Effect of Climate Change on Indoor Air Quality and Environmental Health, provides a response to that charge.

FRAMEWORK AND ORGANIZATION

The first three chapters of the report present introductory and background materials. Subsequent chapters address five major issues related to potential alterations in indoor environmental quality (IEQ) induced by climate change:

- The chemical, organic, and particulate pollutants that can be found in the indoor environment—including infiltrates from the outdoors and pollutants resulting from indoor combustion and other indoor emission sources—and the possible health effects of exposure to them (Chapter 4).
- The health implications of damp indoor spaces, including the effects of exposure to mold and bacteria and their components and to outgassing from the degradation of wet building materials (Chapter 5).
- How various infectious agents, insects, and arthropods that can be found indoors may be affected by climate change (Chapter 6).
- The physiologic, economic, and social factors that influence vulnerability to prolonged exposure to temperature and humidity

extremes and the resources available to mitigate such conditions, including air conditioning and other active and passive means to control the indoor thermal environment (Chapter 7).
- How human health is influenced by building energy use, emissions from building materials, weatherization, and ventilation and possible means to ameliorate adverse effects (Chapter 8).

The sections below are a synopsis of the committee's major findings, conclusions, and recommendations.

REPORT SYNOPSIS

Why the Effect of Climate Change on the Indoor Environment and Health Is an Issue

Indoor environmental conditions exert considerable influence on health, learning, and productivity. Poor environmental conditions and indoor contaminants are estimated to cost the US economy tens of billions of dollars a year in exacerbation of illnesses, allergic symptoms, and lost productivity (Fisk and Rosenfeld, 1997).

Climate change has the potential to affect the indoor environment. There is a large literature on how the indoor environment influences occupant health and how the external environment influences the indoor environment under different climate conditions. Research on the possible effects of climate change on human health is also emerging. However, the intersection of those bodies of research—the fraction specifically on the effects of climate change on human health in the indoor environment—is small. Such studies are complicated by the fact that the effects of climate change on indoor environmental quality are region-dependent and vary with the age and condition of the regionally dependent built environment.

Multiple parts of government and the private sector have a stake in issues of climate change, indoor environmental quality, and public health, but no one body has lead responsibility. As a result, there is a lack of leadership in identifying potential hazards, formulating solutions, and setting research and policy priorities.

Elements of Climate-Change Research Relevant to the Indoor Environment and Health

A 2010 National Academies report concluded that climate change "poses significant risks for a broad range of human and natural systems" (NRC, 2010a). Measurements indicate that the first decade of the 21st century was warmer than the first decade of the 20th century. In the United

States, hot days, hot nights, and heat waves have become more frequent in recent decades. On an urban scale, the heat-island effect contributes to local temperature increases. Rainfall measurements show that extreme events are increasing, moist regions are becoming wetter, and semiarid regions are becoming drier. Projections suggest that those trends will continue and may intensify.

Indoor Air Quality

Three classes of factors have important influences on the indoor concentration of a pollutant: the pollutant's source properties and other attributes, building characteristics, and human behavior. Climate change can affect these factors in numerous ways. Changes in the outdoor concentrations of a pollutant due to alterations in atmospheric chemistry or other factors such as atmospheric circulation will affect indoor concentrations. Mitigation measures to reduce energy use in buildings could lead to systematically lower ventilation rates that would cause higher concentrations and exposures to secondhand smoke and other indoor pollutants. Increased use of air conditioning, an expected adaptation measure, could exacerbate emissions of greenhouse gases and, if accompanied by reduced ventilation rates, increase the concentrations of pollutants emitted from indoor sources. The potential for poisoning from exposure to carbon monoxide emitted from portable electricity generators may increase if peak electricity demand due to heat waves or extreme weather events leads to power outages.

Combustion is a major source of both outdoor and indoor air pollution and is arguably the most important class of indoor air pollutants with respect to health risks. Use of solid-fuel stoves, which are much more common in less developed countries, is associated with demonstrable adverse effects. Switching to lower-emissions units would yield substantial health benefits and decreases in the production of greenhouse gases.

Dampness, Moisture, and Flooding

Studies reviewed in the 2004 IOM report *Damp Indoor Spaces and Health* and confirmed by later research indicates that

- Excessive indoor dampness is a determinant of the presence or source strength of several potentially problematic exposures. Damp indoor environments favor house-dust mites and the growth of mold and other microbial agents, standing water supports cockroach and rodent infestations, and excessive moisture may initiate or increase chemical emissions from building materials and furnishings.

- Damp indoor environments are associated with the initiation or exacerbation of a number of respiratory ailments.

Extreme weather conditions associated with climate change may lead to breakdowns in building envelopes followed by sudden infiltration of water into indoor spaces. Dampness problems and water intrusion create conditions favorable to the growth of fungi and bacteria and may cause building materials to decay or corrode; this can lead to off-gassing of chemicals. Well-designed and properly operating heating, ventilation, and air-conditioning (HVAC) systems can ameliorate humid conditions, but poorly designed or maintained systems may introduce moisture and create condensation on indoor surfaces. Mold-growth prevention and remediation activities may also introduce fungicides and other agents into the indoor environment, which can lead to adverse exposures of occupants.

Infectious Agents and Pests

Weather fluctuations and seasonal to annual climate variability influence the incidence of many infectious diseases. Climate change may affect the evolution and emergence of infectious diseases by, for example, affecting the geographic range of disease vectors. However, relationships between climate and infectious disease often depend heavily on local conditions and may be influenced by indoor characteristics such as air conditioning, which affects indoor temperature and humidity, so it is difficult to draw general conclusions.

The ecologic niches for pests will change in response to climate change. Although decreases in populations in some locations may lower the incidence of allergic reactions to particular pests, the overall incidence of allergic disease may not go down, because those with a predisposition to allergies may become sensitized to other regional airborne allergens.

Climate change may also lead to shifting patterns of indoor exposure to pesticides as occupants and building owners respond to infestations of pests like termites whose geographic ranges have changed.

Thermal Stress

Extreme heat and cold have several well-documented adverse health effects. The elderly, those in poor health, the poor, and those who live in cities are more vulnerable to both exposure to temperature extremes and the effects of exposure. Those populations experience excessive temperatures almost exclusively in indoor environments. Air conditioning provides protection from heat but is associated with higher reported prevalences of some ailments, perhaps because of contaminants in HVAC systems. It also protects against exposure to high concentrations of outdoor pollutants. Temperate indoor conditions are associated with higher work productivity than colder or warmer environments.

Available information on the effects of climate change on building energy use and occupant health indicates that changing conditions may have the following effects:

- Buildings that are currently ventilated naturally will need to use some form of air conditioning.
- Buildings that have air conditioning will need to use it more often, reducing natural ventilation.
- People in buildings that do not have air conditioning will be exposed to extreme heat conditions more often.

Several technologies and building-design and -siting approaches can provide control of the indoor environment with lower energy costs and greater health benefits than systems typically in use today. No matter which approach is used to maintain safe indoor environmental conditions, it is important to ensure that the conditions are sustained when failures in building systems or power outages disable mechanical ventilation—something that may happen more often if climate change leads to more instances of extreme weather conditions or unsustainable loads on the electric grid.

Building Ventilation, Weatherization, and Energy Use

Research indicates that poor ventilation in homes, offices, and schools is associated with occupant health problems or lower productivity. However, the information base is limited, and studies in hot and humid climates are lacking. Climate change may make ventilation problems more common or more severe in the future by stimulating the implementation of energy-efficiency (weatherization) measures that limit the exchange of indoor air with outdoor air.

Introduction of new materials and weatherization techniques may lead to unexpected exposures and health risks. Energy-efficiency programs should therefore incorporate tracking mechanisms to identify problems with indoor environmental quality as they arise and to gather information on the effectiveness of solutions as they are developed and implemented.

Government and consensus organizations are beginning to recognize the importance of this issue and have established or are establishing voluntary guidelines and codes that account for the links between energy efficiency, indoor environmental quality, ventilation, and occupant health and productivity. Problems will persist, however, unless the weatherization workforce is trained to recognize and avoid problems with indoor environmental quality, the efficacy of guidelines and codes is validated, and they are widely implemented.

RESULTS

While there is substantial scientific literature on the effects of outdoor environmental conditions on the indoors, of indoor environmental conditions on health, of climate change on health, of climate change on buildings, and of buildings on climate change, there is almost no literature on the intersection of climate change, indoor environmental quality, and occupant health—and much of what little literature there is summarizes information on one or more of the above categories rather than offering original contributions. The committee was thus required to approach its task by reviewing the available information on components of the climate-change–IEQ–occupant-health nexus and deriving its results on the basis of a synthesis of that information.

The observations and recommendations are based on the committee's review of the scientific literature and on general conclusions reached in previous National Academies reports on climate change and literature those reports found to be authoritative. They do not depend on any particular model of future climatic conditions. The literature on indoor environmental quality and health is rich and unequivocal: indoor environmental conditions have a great influence on human health, and adverse conditions harm occupant well-being. Altered climatic conditions will not necessarily introduce new risks for building occupants but may make existing indoor environmental problems more widespread and more severe and thus increase the urgency with which prevention and interventions must be pursued.

The concluding chapter of the report (9) explicates the key findings, guiding principles, and priority issues for action and recommendations presented below.

Key Findings

Three key findings derived from the committee's literature review underlie its conclusion that alterations in indoor environmental quality induced by climate change are an important public-health problem that deserves attention and action.

Poor indoor environmental quality is creating health problems today and impairs the ability of occupants to work and learn.

There is inadequate evidence to determine whether an association exists between climate-change–induced alterations in the indoor environment

and any specific adverse health outcomes. However, available research indicates that climate change may make existing indoor environmental problems and introduce new problems by

- **Altering the frequency or severity of adverse outdoor conditions that affect the indoor environment.**
- **Creating outdoor conditions that are more hospitable to pests, infectious agents, and disease vectors that can penetrate the indoor environment.**
- **Leading to mitigation or adaptation measures and changes in occupant behavior that cause or exacerbate harmful indoor environmental conditions.**

Opportunities exist to improve public health while mitigating or adapting to alterations in indoor environmental quality induced by climate change.

Guiding Principles

The mission of public health is to "[fulfill] society's interest in assuring conditions in which people can be healthy," and its aim is "to generate organized community effort to address the public interest in health by applying scientific and technical knowledge to prevent disease and promote health" (IOM, 1988). The committee took a public-health approach in formulating its recommendations for reducing the health effects of alterations in IEQ induced by climate change, which can be summarized in three guiding principles:

Prioritize consideration of health effects into research, policy, programs, and regulatory agendas that address climate change and buildings.

As the country moves toward a future where climate change will spur the need for increased action to lower buildings' energy demands and increase their resistance to adverse outdoor conditions, it is vital that public health be put in the forefront of the criteria taken into account in making decisions on issues that affect indoor environmental quality.

Make the prevention of adverse exposures a primary goal when designing and implementing climate change adaptation and mitigation strategies.

Prevention is a foundation principle in public health. Indoor environments already present myriad opportunities for adverse exposures. Common sense suggests that eliminating or lessening those exposures and limiting the introduction of new agents should be the first consideration when responding to potential problems.

Collect data to make better-informed decisions in the future.

A central aim of public-health professionals is "to maximize the influence of accurate data and professional judgment on decision-making—to make decisions as comprehensive and objective as possible" (IOM, 1988). Collecting data that support assessments of the effects of climate change on the indoor environment and health and data on the effects of mitigation and adaptation measures on health will allow future policy to be set in a more informed manner and help to identify misguided or inefficient approaches so that they can be corrected.

Priority Issues for Action and Recommendations

Chapters 4–8 offer several observations regarding how climate change may affect indoor air quality; dampness, moisture, and flooding; infectious agents and pests; exposure to thermal stress; and building ventilation, weatherization, and energy use. The items below constitute a distillation of the committee's thoughts on how their findings and conclusions should be operationalized.

The committee recommends that the Environmental Protection Agency undertake the following actions.

The Environmental Protection Agency should work with such agencies as the Centers for Disease Control and Prevention to assist state, territorial, and local health and emergency-management agencies in efforts to initiate or expand programs to identify populations at risk for health problems resulting from alterations in indoor environmental quality induced by climate change and to implement measures to prevent or lessen the problems.

EPA is a source of expertise on a number of issues related to the indoor environment and health. The Centers for Disease Control and Prevention (CDC)—which has the lead federal role in monitoring health, detecting and investigating health problems, and developing and implementing responses—already works with EPA on topics of common interest, such as the health effects of dampness and mold. Such cooperation will become more important if extreme weather events become more frequent or severe. EPA's knowledge in such fields as weatherization will be of great use in anticipating which future populations may be at risk and in developing solutions.

The committee recommends that interagency collaboration between EPA and CDC expand into emerging issues of climate change and indoor environmental quality. Populations whose health, economic situation, or social circumstances make them more vulnerable to adverse consequences will require special attention in this regard.

> **The Environmental Protection Agency and other federal agencies should join to develop or refine protocols and testing standards for evaluating emissions from materials, furnishings, and appliances used in buildings and to promote their use by standards-setting organizations and in the marketplace. Standards should include consideration of emissions over the operational life of products and the effects of changes in indoor temperature, dampness, and pests.**

Prevention of adverse exposures to materials in the indoor environment and those introduced as a part of weatherization and other climate-change mitigation activities should have high priority, but relatively little information is available. Organizations and government entities in the United States and other countries are pursuing and promoting testing protocols, but these efforts are fragmentary. Facilitating the development of uniform test standards not only will let builders and occupants make more informed decisions about which materials, furnishings, and appliances to use in buildings but will simplify compliance for manufacturers. The committee recommends that EPA pursue expanded and coordinated action with other federal agencies, which will help to ensure that protocols are comprehensive and will promote their acceptance.

> **The Environmental Protection Agency should expand and accelerate its efforts to ensure that indoor environmental quality is protected and enhanced in building-weatherization efforts by facilitating research to identify circumstances in which mitigation and adaptation measures may cause or exacerbate adverse exposures; by reviewing and, where appropriate, changing weatherization guidance to prevent these exposures; and by establishing criteria for the certification of weatherization contractors in health-protective procedures.**

One of the primary points made in this report is that buildings are complex systems whose siting, design, and operation interact in ways that are not necessarily easy to predict. EPA and the Department of Energy (DOE) are already cooperating on protocols for home energy-conservation upgrades that were in draft form when the committee completed its report. Such recognition of health effects on both occupants and persons performing weatherization work is welcome. The committee recommends that it be followed, however, by surveillance activities that evaluate whether guidance is achieving its health-protective objectives and recommends that a mechanism be put into place to revise guidance on the basis of evaluation. It also recommends certification of weatherization contractors in health-protective procedures, which would allow consumers to make better-informed decisions on whom they choose to perform work and give governments and utilities guidance on potential service providers.

The Environmental Protection Agency in coordination with the Department of Energy, the American Society of Heating, Refrigerating and Air-Conditioning Engineers, and building-code organizations should facilitate the revision and adoption of building codes that are regionally appropriate with respect to climate-change projections and that promote the health and productivity of occupants.

EPA works in cooperation with the American Society of Heating, Refrigerating and Air-Conditioning Engineers (ASHRAE), a professional organization, in developing guidelines for indoor air quality and ventilation. DOE works with ASHRAE and other stakeholders on building energy codes. ASHRAE standards for building ventilation and thermal comfort are often incorporated in building codes. The committee recommends that those cooperative efforts on codes be extended to encompass climate-change issues. Most residential and commercial buildings have useful lifetimes that are measured in decades. Promoting research on and development and adoption of regionally appropriate building codes that account for the possibility of future climatic conditions not only will protect the well-being of occupants but could produce economic benefits in the form of longer building lives, lower building insurance fees, and avoided retrofitting costs.

The Environmental Protection Agency and other public agencies and private organizations should join to develop model standards for ventilation in residential buildings and to foster updated standards for commercial buildings and schools. The standards should

- **Be based on health-related criteria.**
- **Account for the effects of weatherization and of other climate-change–related retrofits of existing buildings.**
- **Provide design and operation criteria for mechanical ventilation systems in new construction.**
- **Include consideration of ventilation system hygiene and ventilation effectiveness.**

- **Address how to maintain proper ventilation throughout the life of the system.**
- **Contain "fail-safe" provisions that allow for sufficient air exchange with the outdoors to sustain occupant well-being in the event of ventilation-system breakdown or an extended power outage.**
- **Achieve the objectives mentioned above in an energy- and cost-efficient manner.**

Current ventilation standards are not based on maintaining the health and productivity of occupants and do not account for the potential effects of climate change on building design and operation and on occupant behavior. The committee believes that action should be taken to address this. New ventilation standards should take into account all the considerations listed above. The committee recommends that EPA foster the development and implementation of standards in cooperation with other stakeholders.

The Environmental Protection Agency and other federal agencies should put into place a public-health surveillance system that uses existing environment and health survey instruments to gather information on how outdoor conditions, building characteristics, and indoor environmental conditions are affecting occupant health and on how these change over time.

Lack of general population information on the influence of buildings on occupant health hampers the setting of priorities and the development of effective interventions. The committee believes that it is important to start collecting such data. The ideal surveillance system for assessing how climate change affects indoor environment exposures and related health effects would collect data from across the nation and have this clear focus. However, there are substantial logistical hurdles in mounting such an effort, and its high cost may not be tenable under current federal budget circumstances. The committee therefore recommends that EPA cooperate with its collaborating agencies to identify means for adapting existing environment and health survey instruments to meet the need. It believes that, although challenges exist, it is possible to identify ways to modify and add to existing instruments such as the National Health and Nutrition Examination Survey (NHANES) and Behavioral Risk Factor Surveillance System (BRFSS) to generate useful data and facilitate combining of databases to perform novel analyses.

The Environmental Protection Agency should exercise a strong level of commitment to educate the public on issues of climate change, the indoor environment, and health. Its efforts should

- **Include materials tailored to those involved in the design, construction, operation, and maintenance of buildings and to occupants of single-family and multifamily residences.**
- **Consider differences in geography, building type, age, and setting (city, suburb, and rural area) and in current and possible future climate conditions.**
- **Contain specific advice on actions that will reduce the effects of climate change on the indoor environment and will improve health.**

If adverse effects of climate change are to be prevented, public education and training of professionals will be integral parts of the solution. Education and outreach—especially to those in vulnerable communities and those who provide services to those communities—could have a large role in preventing or limiting adverse effects by making people mindful of potential problems and of the means of addressing them. The committee recommends that EPA expand its current efforts by creating and disseminating specifically tailored messages that speak to the specific circumstances and needs of the diverse audiences listed above and that are focused on steps that these audiences can take to improve indoor environmental quality in the spaces that they occupy.

The Environmental Protection Agency should continuously evaluate actions taken in response to climate-change–induced alterations in the indoor environment to determine whether they are enhancing occupant health and productivity in a cost-effective manner, should identify initiatives that fail to achieve these objectives, and should take corrective steps as needed.

There is little available research on how changes in climatic conditions may affect the indoor environment. It will therefore be especially important to follow up on the measures taken to lessen adverse effects to determine whether they are effective and whether there are more efficient means of achieving the desired outcomes. The committee therefore recommends that intervention programs include the collection of data that will allow evaluation of whether the programs are materially affecting the health of occupants.

The Environmental Protection Agency should spearhead an effort across the federal government to make indoor environment and health issues an integral consideration in climate change research and action plans and, more broadly, to coordinate work on the indoor environment and health.

The serious gap in the scientific literature concerning the relationships among climate change, IEQ, and occupant health identified in this report is a barrier to effective action on the issue. In the committee's judgment, there is a clear lack of recognition of this topic at a level commensurate with its importance.

At the US federal level, the research gap is emblematic of a more fundamental problem regarding indoor environmental health concerns: that responsibility for the integrated environmental, public-health, energy-conservation, housing, urban-planning, and worker well-being issues that make up IEQ do not fall neatly under the aegis of any federal department or agency. Because several organizations have interests in some subjects, yet no entity has the lead responsibility, research needs go unrecognized and unmet, and opportunities for efficient action are unrealized.

The committee believes that this situation must change. Several of the priority issues listed above recommend that EPA either initiate or deepen their cooperation with governmental and other entities on some specific urgent issues and achievement of their goals will be predicated on building and sustaining robust partnerships. The committee believes that these initiatives should be part of a larger effort to entwine indoor environment and health considerations into the fabric of research and action plans. As it is difficult to separate the effects of climate change from other influences on the indoor environment, a broad approach to IEQ issues is needed.

There are several potential approaches to addressing the problem.

One is for EPA to initiate action within the US Global Change Research Program (USGCRP)—in which it participates—to address the effects of climate change on indoor environmental quality and on the health and productivity of occupants. The USGCRP, which involves 13 federal departments and agencies, serves as the coordinating body for federal research on climate change and its effects on society (CCHHG, 2011). The USGCRP is in the process of formulating a new strategic plan with the intent of releasing it in December, 2011. This process presents an opportunity for EPA to advocate for the inclusion of indoor environment and health concerns into the work of the Program and in particular, the adaptation science; assessments; and communication, education, and engagement elements of the new strategic plan.

EPA should also explore options for stimulating action on climate change, indoor environment, and health issues outside and within the government. These include the initiatives highlighted in the committee's recommendation above that the agency exercise a strong level of commitment to educate the public on these issues.

At the federal level, the committee suggests that EPA promote a broader coordinated effort to address indoor environment and health issues through, for example, the establishment of an interagency working group or a national center. Such mechanisms have been used to effectively coordinate action to identify information gaps, facilitate research, collect data, and catalyze work on other critical issues. An effort to establish a governmental entity to act as a coordinating body will likely require support from the administration or Congress. Nonetheless, the committee believes that consolidating and focusing indoor environmental health efforts may generate efficiencies that make it worthy of consideration and that any efforts that support collaboration in the pursuit of healthy indoor environments will produce societal benefits.

The United States is in the midst of a large experiment of its own making in which weatherization efforts, energy-efficiency retrofits, and other initiatives that affect the characteristics of interaction between indoor and outdoor environments are taking place and new building materials and consumer products are being introduced indoors with little consideration of how they might affect the health of occupants. Experience provides a strong basis to expect that some of the effects will be adverse, a few profoundly so. An upfront investment in considering the consequences of these actions before they play out and thereby avoiding problems that can be anticipated would yield benefits in health and in avoiding costs of medical care, remediation, and lost productivity.

REFERENCES

CCHHG (Interagency Crosscutting Group on Climate Change and Human Health). 2011. *Interagency Crosscutting Group on Climate Change and Human Health*. http://www.globalchange.gov/what-we-do/climate-change-health (accessed February 27, 2011).

Fisk WJ, Rosenfeld AH. 1997. Estimates of improved productivity and health from better indoor environments. *Indoor Air* 7:158-172.

IOM (Institute of Medicine). 1988. *The future of public health*. Washington, DC: National Academy Press.

IOM. 2004. *Damp indoor spaces and health*. Washington, DC: The National Academies Press.

NRC (National Research Council). 2010a. *Advancing the science of climate change*. Washington, DC: The National Academies Press.

NRC. 2010b. *Informing an effective response to climate change*. Washington, DC: The National Academies Press.

1

Introduction

This chapter provides basic information about the report's motivation and the conduct of the study, beginning with an overview of why the effects of climate change on the indoor environment and health constitute an important issue. It then presents the statement of task for the Institute of Medicine (IOM) committee responsible for this report, which is followed by the committee's approach to its task. The text then addresses some of the methodologic considerations that informed the committee's evaluation of the literature and concludes with a description of the report's organization.

WHY THE EFFECT OF CLIMATE CHANGE ON THE INDOOR ENVIRONMENT AND HEALTH CONSTITUTES AN IMPORTANT ISSUE

The indoor environment affects comfort, health, and productivity. People in developed countries spend most of their time indoors, so most of the adverse exposures that they encounter regularly take place indoors. Many exposures that are potentially hazardous to health are exposures to substances emitted indoors from indoor sources. Such emissions can occur from building materials; from products used or stored indoors; from processes that occur in indoor environments; from the microorganisms, insects, other animals, and plants that live indoors; and from the behavior of building occupants. Because of the contributions from indoor sources, indoor levels of many pollutants are higher than those found outdoors. In addition to pollutants attributable to indoor sources, ventilation may draw pollutants into buildings from outdoor air. Buildings offer protection against some pollutants that are of predominantly outdoor origin; but that protection is generally incomplete. And some outdoor pollutants that enter a building interact with its components or contents and thereby alter the composition of indoor air in ways that can affect the health and welfare of occupants.

Climate change has the potential to affect the indoor environment. Ambient conditions in the outdoor environment serve as boundary conditions to the ambient conditions of the indoor environment. Outdoor air temperature, humidity, air quality, precipitation, and land surface wetness can all influence the indoor environment, depending on such factors as the integrity of a building's envelope; the state of its heating, ventilation, and air-conditioning systems; the inhabitants of the outdoor ecosystem; and the characteristics of the buildings around it. If climatic conditions in a particular area change—for example, if the climate becomes warmer or if there are more severe or more frequent episodes of high heat or intense precipitation—buildings (and other infrastructure) that were designed to operate under the "old" conditions may not function well under the "new." Furthermore, in responding to climate changes, people and societies will seek to mitigate undesirable changes and adapt to changes that cannot be mitigated. Some of their responses will play out in how built spaces are designed, constructed, used, maintained, and in some cases retrofitted, and the actions taken may well have consequences for indoor environmental quality and public health.

There is a body of literature on how the indoor environment influences occupant health and how the external environment influences the internal built environment under past and present climate conditions. And research is emerging on the possible effects of climate change—such as extreme temperatures and thermal stress, vectorborne infectious diseases, and outdoor air quality—on human health. However, the body of research specific to the effects of climate change on human health in the indoor environment is very small. Such studies are complicated by the fact that the effects of climate change on, say, indoor air quality depend on the geographic region and are a function of the age and condition of the regionally dependent built environment.

Against that backdrop, the US Environmental Protection Agency (EPA) approached IOM with a request to summarize and benchmark the state of the science concerning the health effects of climate change–induced alterations in the indoor environment, raise awareness of crucial issues, and suggest a way forward. The Committee on the Effect of Climate Change on Indoor Air Quality and Public Health was formed to respond to that request.

STATEMENT OF TASK

EPA charged the committee to develop a report summarizing the current state of scientific understanding of the effects of climate change on indoor air and public health. It provided three examples of key questions to address:

- What are the likely impacts of climate change in the United States on human exposure to chemical and biological contaminants inside buildings, and what are the likely public health consequences?
- What are the likely impacts of climate change on moisture and dampness conditions in buildings, and what are the likely public health consequences?
- What are priority issues for action?

EPA indicated that it intended the report to serve as the foundation for the development of US government funding priorities and for use in communications to and guidance for the public.

THE COMMITTEE'S APPROACH TO ITS TASK

To answer the questions posed by EPA, the committee undertook a wide-ranging evaluation of relevant research on climate change, buildings, indoor environmental quality, and occupant health. Although the committee did not review all such literature—an undertaking beyond the scope of this report—it did attempt to cover the work that it believed to have been influential in shaping scientific understanding by at the time it completed its task in early 2011.

The committee consulted several sources of information. On health outcomes, the primary source was epidemiologic studies. Most of those studies examined general population exposures to problematic agents in homes, reflecting the focus of researchers working in the field. The committee also examined the smaller literature addressing commercial buildings, apartments, schools, and other buildings. Clinical and toxicologic research were considered as appropriate.

The literature of engineering, architecture, and the physical sciences informed the committee's discussions of building characteristics, exposure assessment and characterization, pollutant transport, and related topics; and public-health and behavioral-sciences research was consulted for the discussion of public-health implications. Those disciplines have different practices regarding the publication of research results. For example, relatively few papers in the peer-reviewed literature address building construction or maintenance issues. The committee endeavored in all cases to identify, review, and consider fairly the literature most relevant to the topics that it was charged to address.

Papers and reports reviewed in this volume were identified through extensive searches of relevant databases. Most were bibliographic and provided citations of peer-reviewed scientific literature. Committee staff examined the reference lists of major papers, books, and reports for relevant citations, and committee members independently compiled lists of potential citations on the basis of their expertise. The input received in both written and oral form from participants at three public meetings held in February–July 2010 served as a valuable source of additional information. Appendix A lists the participating researchers and their topics.

The committee also relied on the research and conclusions of prior National Academies committees that addressed indoor environment and health issues. The 2004 IOM report *Damp Indoor Spaces and Health* and the 2006 National Research Council report *Green Schools: Attributes for Health and Learning* (NRC, 2006) were particularly influential. Research published after their completion dates is used to supplement this material.

The committee did not attempt to review and evaluate the literature regarding potential effects of climate change on the outdoor environment or health independently. Several National Academies reports have addressed those topics in detail, including *Global Climate Change and Extreme Weather Events: Understanding the Contributions to Infectious Disease Emergence* (NRC, 2008) and four published in 2010: *Advancing the Science of Climate Change* (NRC, 2010b), *Limiting the Magnitude of Climate Change* (NRC, 2010d), *Adapting to the Impacts of Climate Change* (NRC, 2010a), and *Informing an Effective Response to Climate Change* (NRC, 2010c). Salient findings, conclusions, and recommendations from those and other National Academies reports are referenced throughout the present report.

EPA also commissioned several white papers addressing various issues related to climate change, the indoor environment, and health to serve as information resources for the committee. The papers, which are listed in Appendix C, were helpful sources of references and perspectives for the committee to consider. In some cases, they delve into topics at a greater level of detail than is present in this report. The papers are the work product of their authors and do not necessarily represent the committee's point of view.

METHODOLOGIC APPROACH

This section presents the general considerations regarding climate change, the indoor environment, and public health that informed the committee's approach to evaluating the scientific literature. It discusses, in general terms, the major issues involved in determining environmental conditions in buildings and how building characteristics, occupant behavior, and the outdoor environment may affect them. The committee's statement of task directed it to focus on indoor air quality (IAQ), a major component of indoor environmental quality (IEQ),[1] and the text reflects that guidance.

General Considerations

As detailed later in this report, little in the literature considers together the key elements in the committee's charge: the effects of climate change on IEQ that would influence public health. However, substantial research has been published on many key questions. For example, there is a strong emerging literature on the effects of climate change on outdoor air pollution. A voluminous literature characterizes health risks associated with pollutants[2] in outdoor air. Considerable published research documents our understanding of indoor–outdoor relationships with respect to important air pollutants. Research has explored the extent to which health risks associated with outdoor pollution are a consequence of indoor exposures. There is a large body of work reporting on how indoor pollution sources influence IAQ and human health, including several National Academies reports (IOM, 1993, 2000, 2004; NRC, 1981). A number of papers are available on the determinants of exposure to indoor dampness and on the association of dampness or dampness-related agents with health outcomes. And the health effects associated with prolonged exposure to temperature extremes is relatively well studied.

However, little published research links climate change to changes in levels of indoor air pollutants or to other changes in indoor environmental conditions that might influence public health. Among the available studies, Ayres et al. (2009)—summarizing how climate change is expected to affect respiratory health—called for more research on "the role of housing and indoor climate control systems in respiratory diseases." Bell et al. (2009) used an epidemiologic approach to discern that communities with higher air-conditioner prevalence exhibited "lower health effects estimates" associated with outdoor particulate-matter levels. The use of air conditioning for residential climate control would be expected to provide better protection against outdoor particles than would opening windows. Peden and Reed (2010) review the many ways in which indoor pollution and outdoor pollution influence the prevalence and severity of allergic diseases. They discuss

[1] Indoor environmental quality is defined by a building's indoor air quality and the comfort of its occupants, which is influenced by factors such as the building's ventilation, temperature, humidity, sound, and light levels.

[2] A pollutant is anything that, at some concentration or level, is harmful to humans or the environment. It includes biologic, chemical, and particulate agents.

the role that climate change will have in altering the spatial and temporal patterns of outdoor aeroallergens. In perhaps the most directly relevant study, Wilkinson et al. (2009) evaluated cobenefits of mitigating climate change and improving public health that would result from improving the residential building stock in the United Kingdom and from an improved stoves program in India.

Even though the climate-change–IEQ–public-health nexus has not yet been well studied, the elements are sufficiently well understood to permit the committee to conduct a scientific examination of issues, come to findings, draw conclusions, and offer recommendations. The approach taken is to identify exposures and exposure circumstances believed to affect the health, safety, or productivity of building occupants; to describe the factors that influence exposure or source strength; and to explore how climate change might influence these factors. Because the analysis relies on inference, the committee was constrained to focus on portions of the system that are well understood mechanistically. In extrapolating from available evidence to explore an unknown future, the committee is on more solid ground when inferences are based on a cause–effect understanding of the system rather than when it has to rely on studies that base associations on statistical methods without providing clear evidence on processes. Because of those limitations, the report stresses how climate-change phenomena might induce changes in adverse exposures. In a few cases, the mechanistic level of understanding is sufficient to relate potential changes in future exposures to health consequences.

Framing the Issues

Fundamentally, exposures occur when people and pollutants intersect in space and time. The magnitude of an exposure depends on its level while a subject is present. Three classes of factors govern conditions in occupied indoor environments. The first pertains to the adverse exposures themselves and includes such factors as the outdoor level and, in some cases, the physical properties of the agent. The second category pertains to buildings and includes the air-exchange rate, the characteristics of temperature and humidity controls, the presence and effectiveness of deliberate air-cleaning processes, and the types and conditions of materials that make up the building surfaces and furnishings; this category also includes factors that affect emissions from materials associated with the building and its (nonhuman) contents. The third category of factors pertains to occupants and includes the timing of their presence indoors, occupant density, and activities that may influence both sources and exposure. Each category is complex: adverse exposures, buildings, and people are both numerous and diverse with regard to many attributes. The factors in each category can influence IEQ and its public-health consequences.

It is convenient to decompose the analysis of indoor exposures into two components: outdoor and indoor sources. For many pollutants, these two components do not interact directly, and the total indoor burden can be represented as their arithmetic sum.[3]

The ventilation or air-exchange rate of a building or of a room in a building can substantially influence indoor air-pollutant concentrations and other environmental conditions. Ventilation is the means by which pollutants of outdoor origin are introduced into an indoor environment. Whether a pollutant is of outdoor or indoor origin, ventilation is commonly an important removal mechanism that limits its accumulation indoors. In fact, a main purpose for ventilating buildings is to remove indoor-generated pollutants, including those emitted by human occupants. In general, higher ventilation rates cause indoor environmental quality to become more like local outdoor environmental quality. Conversely, as ventilation rates are reduced, the indoor environment is progressively less influenced by pollutants of outdoor origin and outdoor environmental conditions and more strongly influenced by indoor sources and conditions.

Climate change could influence IEQ in many ways. First, considering the existing building stock, a substantial influence can be expected from

- Changes in the levels of outdoor air pollutants or other outdoor conditions, which affect indoor human exposure from outdoor sources.
- Changes in how buildings are operated, for example, with respect to ventilation rate or air-conditioner use, which in turn alters indoor conditions.
- Adjustments in how occupants behave—for example, changing where they spend time or what they do indoors—in response to outdoor conditions and the resulting changes in the indoor environment or in exposure opportunities.

Climate-change effects may occur over decades and one should expect concomitant changes in the building stock. These building-stock changes might substantially influence the nature of climate change and its effects on IEQ and health. There might also be changes in how occupants behave in buildings that evolve on decadal time scales and materially alter the level and nature of indoor exposures.

[3] An example of this approach in the case of particulate matter—specifically, the mass concentration of particles finer than 10 μm in diameter, that is, PM10—is given by Ott et al. (2000).

A change in building design, building operation, or habitual indoor human behavior that is influenced by climate change might be categorized as either an *adaptation* or a *mitigation*. An adaptation is a change made in response to climate change to provide protection against its effects. Increased use of air conditioning would be an adaptation in response to higher average ambient temperatures. Mitigation is a change made to reduce or offset an effect. Because a large proportion of society's use of fossil fuels is associated with buildings, buildings are and will probably continue to be settings where improved energy performance is sought. Some changes motivated by the goal of saving energy can have consequences for IEQ and public health.

In addition to adaptation and mitigation that can be expected, one should be mindful of behavioral responses to climate catastrophes that may themselves have serious consequences for IEQ and public health. Examples would be actions taken to protect people and property in response to floods, extreme heat events, or power outages. A specific concern that is discussed in more detail later in this report is the indoor use of back-up electricity generators after extreme weather events, which has been associated with carbon monoxide (CO) poisonings (Hampson and Stock, 2006).

The effects of climate change on IEQ will probably depend on building type. The consequences of the effects will depend on how long people spend in different types of indoor environments and on differences in the populations that occupy various building types. As detailed in Chapter 2, people spend most of their time in their own residences. Children spend a high proportion of their time in school, and they are considered more vulnerable than adults to adverse health effects of air pollution. Analogously, indoor environments occupied by the elderly or where health care is provided would be of special concern because those who are in fragile health are more vulnerable to further stresses than those who are healthy.

Differentiating among building types is important for reasons that extend beyond the populations that inhabit them. Different classes of buildings may be designed, operated, and maintained differently in ways that affect their responsiveness to climate change. For example, office buildings in the United States are commonly ventilated mechanically whereas the existing stock of residential buildings is ventilated mainly by a combination of air leakage (infiltration) and natural ventilation through open windows or doors. Buildings also differ in types of pollutant-emitting sources of concern. For example, cooking is a dominant activity in restaurants, common in residences, and rare in offices. Candle use is largely confined to restaurants and in residences. The intensity of use of cleaning products may be higher in health-care facilities than in other types of buildings. Finally, it is important to recognize that the responsibility for environmental conditions in buildings varies markedly among building classes and that this variability influences the appropriateness of policy options to address the public-health concerns discussed here.

Another important characteristic of indoor environments is their broadly distributed nature. That results in far greater diversity in indoor environmental conditions than tends to occur outdoors. Consider, for example, that in the United States, more than half the population lives in the 52 most populous metropolitan statistical areas (MSAs), as defined by the Office of Management and Budget. Although there is some local and neighborhood variability in air-pollutant concentrations in those areas, there are also some common characteristics, and the air quality of each MSA can be reasonably characterized by using a relatively small number of monitoring stations. Furthermore, the actions of small numbers of individuals in an MSA have little influence on urban air quality. In contrast, the population of the United States resides in about 100 million residential units, and there are tens of millions of other occupied buildings in the US stock. What happens in individual buildings strongly influences the quality of the indoor environment in those buildings but generally does not substantially affect IEQ in other buildings.

In turn, the IEQ in a given building can affect the health of people occupying that building but generally would not affect others. Diversity in building stock is especially important for understanding the public-health significance of how climate change might affect IEQ. Subpopulations that are potentially vulnerable to the adverse consequences of climate-change–induced effects on IEQ include not only those who are more susceptible to air-pollutant health effects or to temperature extremes because they are young, old, or infirm but those who lack the financial resources or the appropriate knowledge to act wisely in response to an emergency induced by a climate-change event.

In light of that broad diversity, what factors affect indoor pollutant levels? According to the principle of material balance (that is, that mass is conserved), the level of a given pollutant in a particular building can be determined by accounting for the net effect of the source terms and the removal processes. Sources include outdoor air and direct indoor emissions. Similarly, indoor dampness and temperature levels are a function of indoor and outdoor levels. Ventilation is a removal process that must always be considered. For some pollutants and for some buildings, other removal processes can be important, such as deposition of particles onto indoor surfaces, irreversible reaction of a pollutant with an indoor surface, or active filtration.

Buildings are ventilated so that the replacement time of indoor air with outdoor air occurs on a time scale that is typically a few hours but may range from about 5 min, in the case of a mechanically ventilated building using an economizer or a building with open doors and windows, to about

10 h, in the case of a closed building that is on the tight end of the normal range. Dynamic, time-dependent relationships governing the relationship between indoor and outdoor levels are important for time scales similar to or shorter than the ventilation time scale, but the time-dependent processes are not as important for evaluating longer-term average conditions. In many epidemiologic studies, consideration of the effects of outdoor on indoor conditions is based on one-time measurements or time-averaged conditions rather than short-term dynamics. However, short-term dynamics are important in the event of high exposure concentrations that lead to acute and severe health effects.

Changes in IEQ can be expected if homes become more tightly sealed as a response to increasing temperatures and humidity outdoors or because of efforts to reduce building energy use. Tightly sealed buildings tend to have decreased ventilation rates and higher levels of indoor-emitted pollutants.

In general, the key elements that help to ensure good IEQ are indoor source control; adequate ventilation; and proper management of indoor environmental conditions through temperature and humidity control and, where appropriate, the use of air filtration, air cleaning, or other mechanisms to achieve further improvements. The central principle is to remove pollutants where they are more highly concentrated, to supply clean air where people need it, and to maintain comfortable environmental conditions for building occupants. The use of exhaust fans in bathrooms and the use of range hoods above cooking appliances, for example, are practical illustrations of efficient ventilation. Deliberate air cleaning for indoor environments is widely practiced only in the case of particle filtration in mechanically ventilated buildings, and there are opportunities to do more.

Chapters 4–8 discuss how indoor environmental conditions might be influenced by climate change. They are not intended to constitute a comprehensive review of the literature but rather to be broadly illustrative of important IEQ concerns that might be influenced by climate change. Most of what follows is concerned with conditions in buildings of the types commonly found in the United States, but the report also addresses an important international public-health problem: exposure to smoke from the indoor combustion of solid biomass and coal, which occurs predominantly in developing countries.

RECENT NATIONAL ACADEMY OF SCIENCES REPORTS ADDRESSING CLIMATE CHANGE

In 2007, the Congress tasked the National Oceanic and Atmospheric Administration to contract with the National Academy of Sciences to

> investigate and study the serious and sweeping issues relating to global climate change and make recommendations regarding what steps must be taken and what strategies must be adopted in response to global climate change, including the science and technology challenges thereof. (Public Law 110-161, §114)

The National Research Council initiated the *America's Climate Choices* research effort in response. This program has produced several publications that offer a broader perspective on climate change issues than is provided in this report. Primary publications are summarized below.[4]

Limiting the Magnitude of Climate Change (NRC, 2010f) describes, analyzes, and assesses strategies for reducing the net future human influence on climate, including both technology and policy options. The report focuses on actions to reduce domestic greenhouse gas emissions and other human drivers of climate change, such as changes in land use, but also considers the international dimensions of limiting climate change.

Adapting to the Impacts of Climate Change (NRC, 2010a) evaluates strategies to adapt to climate change in different regions, sectors, systems, and populations. The report reviews options and barriers to reduce vulnerability; increase adaptive capacity; improve resiliency; and promote successful adaptation. This report identifies lessons learned from past experiences, promising current approaches, and a framework for a national adaptation strategy.

Advancing the Science of Climate Change (NRC, 2010b) provides an overview of past, present, and future climate change, including its causes and its impacts; and recommends steps to advance our current understanding, including new observations, research programs, next-generation models, and the physical and human assets needed to support these and other activities. The report focuses on the scientific advances needed both to improve the understanding of the integrated human-climate system and to devise more effective responses to climate change.

Informing an Effective Response to Climate Change (NRC, 2010e) describes and assesses different activities, products, strategies, and tools for informing decision-makers about climate change and helping them plan and execute effective, integrated responses. The report describes the different types of climate change-related decisions and actions being taken at various levels and in different sectors and regions; and develops a framework, tools, and practical advice for ensuring that the best available technical knowledge about climate change is used to inform these decisions and actions.

America's Climate Choices (NRC, 2011), the final report in the series,

[4] The summaries below are adapted from descriptions contained in NRC, 2010a.

recommends actions that should be taken at the national level to minimize the risks associated with climate change. It proposes an iterative risk management approach that comprises "identifying risks and response options, advancing a portfolio of actions that emphasize risk reduction and are robust across a range of possible futures, and revising responses over time to take advantage of new knowledge." The report also recommends a coordinated effort across the government to conduct research on adaptation and other climate change issues.

Among these, *Advancing the Science of Climate Change* addresses the issues most closely related to this report. Although it does not mention the indoors specifically, it does devote chapters to both public health and cities and built environment, and briefly touches on energy efficiency improvements. The key research needs identified by the study include the following:

- Characterize the differential vulnerabilities of particular populations to climate-related impacts, and the multiple stressors they already face or may encounter in the future.
- Identify effective, efficient, and fair adaptation measures to deal with health impacts of climate change.
- Develop integrated approaches to evaluate ancillary health benefits (and unintended consequences) of actions to limit or adapt to climate change.
- Develop and test approaches for limiting and adapting to climate change in the urban context, including, for example, the efficacy of and social considerations involved in adoption and implementation of white and green roofs, landscape architecture, smart growth, and changing rural-urban socioeconomic and political linkages.
- Improve understanding of urban governance capacity, and develop effective decision support tools and approaches for decision making under uncertainty, especially when multiple governance units may be involved.
- Develop better understanding of informing, communicating with, and educating the public and health professionals as an adaptation strategy.

In addition, two 2010 workshop reports from the National Research Council contain relevant information. *Facilitating Climate Change Responses* (NRC, 2010d) illustrates some of the ways the behavioral and social sciences can contribute to climate research. It addresses both mitigation—which it defines as "behavioral elements of a strategy to reduce the net future human influence on climate"—and adaptation—"behavioral and social determinants of societal capacity to minimize the damage from climate changes that are not avoided"—strategies, and includes discussions of the ways to stimulate behavioral changes that achieve emissions reductions from household actions and induce household investments in energy efficiency.

Describing Socioeconomic Futures for Climate Change Research and Assessment (NRC, 2010c) notes that the implications of climate change for the environment and society depend not only on the rate and magnitude of climate change, but also on changes in technology, economics, lifestyles, and policy that affect the capacity both for limiting and adapting to climate change. The report explores driving forces and key uncertainties that affect impacts, adaptation, vulnerability, and mitigation and considers research needs and the elements of a strategy for describing socioeconomic and environmental futures for climate change research and assessment.

ORGANIZATION OF THE REPORT

The remainder of this report is divided into eight chapters and supporting appendixes. Chapter 2 sets the scene for the later sections by providing background information on a set of topics relevant to the consideration of the intersections of climate change, the indoor environment, and public health. They include the elements of climate-change research most relevant to the indoor environment, how the outdoor environment affects conditions indoors, how the indoor environment affects health, and the amount of time that people spend indoors. The chapter also addresses populations that are particularly vulnerable to health problems associated with the indoor environment. It identifies the five major issues related to potential alterations in IEQ induced by climate change: air quality; dampness, moisture, and flooding; infectious agents and pests; thermal stress; and building ventilation, weatherization, and energy use.

Several government and private-sector bodies are involved in various aspects of issues of climate change, the indoor environment, and health issues. Chapter 3 identifies them and summarizes their work. It also lists some major sources of data on the characteristics of buildings, the indoor environment, and health, and discusses how they might inform questions about the intersection between these three topics.

Chapter 4 examines the first of the report's major issues: indoor air quality. It focuses on the sources and health effects of chemical and particulate pollutants that can be found suspended in air and in some cases deposited on or sorbed to indoor surfaces. The text addresses volatile and semivolatile molecular pollutants, both organic and inorganic, and abiotic particulate matter. There are also brief discussions of allergens associated with pollen, of respiratory health risks associated with algal blooms after floods, and of CO exposure associated with the use of home electricity generators typically used during power outages. The chapter concludes with a discussion of an important international public-health problem: exposure in developing countries to smoke from the indoor combustion of solid biomass and coal.

IEQ problems associated with dampness, moisture, and flooding are addressed in Chapter 5. The problems include the effects of exposure to mold and hydrophilic bacteria and their components and exposure to degradation products of wet materials. The discussion in this chapter builds on a set of major literature reviews, including the IOM report *Damp Indoor Spaces and Health* (IOM, 2004), highlighting their findings and other research relevant to the consideration of the health effects of alterations in IEQ induced by climate change.

Chapter 6 addresses IEQ concerns associated with infectious agents, insects and arthropods, and mammals that research suggests may be influenced by climate-change–induced alterations in the indoor environment. The chapter also touches on exposures to chemicals used to control pest infestations in buildings.

"Thermal Stress," Chapter 7, considers IEQ problems associated with the thermal environment of buildings, how climate change could induce alterations in the frequency or severity of problems, and some of the means available to mitigate adverse conditions. Thermal stress is a particular threat to certain populations whose health, economic situation, or social circumstances make them vulnerable to exposure to temperature extremes or the consequences of such exposure, and the text thus focuses on these groups. Because climate models suggest that trends toward longer and more extreme heat waves and shorter and milder cold spells will continue and intensify, much of the information presented in the chapter relates to issues involving prolonged exposure to high temperature.

Chapter 8 concludes the discussion of major issues related to potential alterations in IEQ induced by climate change. It focuses on building energy use, emissions from building materials, weatherization and ventilation, and how these affect occupants. The chapter includes the topics of energy consumption in buildings, the means used to tighten buildings, programs to enhance the energy efficiency of buildings and reduce harmful emissions from building components, the training of personnel who implement weatherization programs, and the effects of tightening on ventilation, IEQ, and occupant health and productivity.

The final chapter of the report—Chapter 9—builds on the foundation of the foregoing to draw out the overarching themes of the report and present the committee's key findings, guiding principles, and high-priority issues for action.

Agendas of the public meetings held by the committee are provided in Appendix A. Appendix B contains summaries of the contents of a set of white papers on topics related to climate change, the indoor environment, and health that were commissioned by EPA to provide information for the committee's consideration. Biographic information on the committee members and staff responsible for this study are provided in Appendix C.

REFERENCES

Ayres JG, Forsberg B, Annesi-Maesano I, Dey R, Ebi KL, Helms PJ, Medina-Ramón M, Windt M, Forastiere F. 2009. Climate change and respiratory disease: European Respiratory Society position statement. *European Respiratory Journal* 34:295-302.

Bell ML, Ebisu K, Peng RD, Dominici F. 2009. Adverse health effects of particulate air pollution: Modification by air conditioning. *Epidemiology* 20:682-686.

Hampson NB, Stock AL. 2006. Storm-related carbon monoxide poisoning: Lessons learned from recent epidemics. *Undersea & Hyperbaric Medicine* 33(4):257-263.

IOM (Institute of Medicine). 1993. *Indoor allergens. Assessing and controlling adverse health effects*. Washington, DC: National Academy Press.

IOM. 2000. *Clearing the air. Asthma and indoor air exposures*. Washington, DC: National Academy Press.

IOM. 2004. *Damp indoor spaces and health*. Washington, DC: The National Academies Press.

NRC (National Research Council). 1981. *Indoor pollutants*. Washington, DC: National Academy Press.

NRC. 2006. *Green schools: Attributes for health and learning*. Washington, DC: The National Academies Press.

NRC. 2008. *Global climate change and extreme weather events: Understanding the contributions to infectious disease emergence: Workshop summary*. Washington, DC: The National Academies Press.

NRC. 2010a. *Adapting to the impacts of climate change*. Washington, DC: The National Academies Press.

NRC. 2010b. *Advancing the science of climate change*. Washington, DC: The National Academies Press.

NRC. 2010c. *Describing socioeconomic futures for climate change research and assessment: Report of a workshop*. Washington, DC: The National Academies Press.

NRC. 2010d. *Facilitating climate change responses: A report of two workshops on insights from the social and behavioral sciences*. Washington, DC: The National Academies Press.

NRC. 2010e. *Informing an effective response to climate change*. Washington, DC: The National Academies Press.

NRC. 2010f. *Limiting the magnitude of climate change*. Washington, DC: The National Academies Press.

NRC. 2011. *America's climate choices*. Washington, DC: The National Academies Press.

Ott W, Wallace L, Mage D. 2000. Predicting particulate (PM_{10}) personal exposure distributions using a random component superposition statistical model. *Journal of the Air & Waste Management Association* 50(8):1390-1406.

Peden D, Reed CE. 2010. Environmental and occupational allergies. *Journal of Allergy and Clinical Immunology* 125:S150-S160.

Wilkinson P, Smith KR, Davies M, Adair H, Armstrong BG, Barrett M, Bruce N, Haines A, Hamilton I, Oreszczyn T, Ridley I, Tonne C, Chalabi Z. 2009. Public health benefits of strategies to reduce greenhouse-gas emissions: Household energy. *Lancet* 374:1917-1929.

2

Background

This chapter provides background information on several topics relevant to the consideration of the intersections of climate change, the indoor environment, and public health. They include the elements of climate-change research most relevant to the indoor environment, how the outdoor environment affects conditions indoors and how the indoor environment affects health, and the amount of time that people spend indoors. The chapter identifies the five major issues related to potential alterations in indoor environmental quality induced by climate change: air quality, dampness, moisture and flooding, infectious agents and pests, thermal stress, and building ventilation, weatherization, and energy use. It also addresses populations that are particularly vulnerable to health problems associated with indoor environmental quality.

ELEMENTS OF CLIMATE-CHANGE RESEARCH RELEVANT TO BUILDINGS AND PUBLIC HEALTH

The science of climate change is large and complex, and many details are outside the scope of the committee's task. It therefore did not conduct an independent review of the voluminous literature regarding such subjects as the nature of changes in the earth's climate in the short and long term and the potential magnitude of the changes. Instead, the committee drew on the research and conclusions contained in other National Academies reports—in particular, four in the *America's Climate Choices* series (NRC, 2010a,b,c,d)—and peer-reviewed literature and assessments found to be authoritative by the committees responsible for those reports, such as the

Intergovernmental Panel on Climate Change Fourth Assessment Report (IPCC, 2007) and *Global Climate Change Impacts in the United States* (USGCRP, 2009).

The overall conclusion of the National Academies report *Advancing the Science of Climate Change* was that climate change "poses significant risks for—and in many cases is already affecting—a broad range of human and natural systems" (NRC, 2010b, p. 1). The US Global Change Research Program, which coordinates and integrates federal climate change research, found (USGCRP, 2009, p. 9) that

> Climate-related changes have already been observed globally and in the United States. These include increases in air and water temperatures, reduced frost days, increased frequency and intensity of heavy downpours, a rise in sea level, and reduced snow cover, glaciers, permafrost, and sea ice. A longer ice-free period on lakes and rivers, lengthening of the growing season, and increased water vapor in the atmosphere have also been observed. Over the past 30 years, temperatures have risen faster in winter than in any other season, with average winter temperatures in the Midwest and northern Great Plains increasing more than 7°F. Some of the changes have been faster than previous assessments had suggested.
>
> These climate-related changes are expected to continue while new ones develop. Likely future changes for the United States and surrounding coastal waters include more intense hurricanes with related increases in wind, rain, and storm surges (but not necessarily an increase in the number of these storms that make landfall), as well as drier conditions in the Southwest and Caribbean. These changes will affect human health, water supply, agriculture, coastal areas, and many other aspects of society and the natural environment.

Such findings are relevant to the committee's work because conditions in the outdoor environment greatly influence conditions in the indoor environment.

Literature Regarding Observations of Climate Change

This report uses the term *climate* to refer to prevailing outdoor environmental conditions—including temperature, humidity, wind, precipitation, sea level, and other phenomena—and *climate change* to refer to modifications in those outdoor conditions that occur over an extended period of time. Observations of key climatic variables provide a rich historical record of how the climate has changed in the past and serve as a basis for assessing potential future change (IPCC, 2007; NRC, 2010b; USCCSP, 2008).

Measurements of global mean temperature indicate that the first decade of the 21st century was 0.8°C (1.4°F) warmer than the first decade of the

20th century. Associated with that temperature rise have been observations that heat waves have become longer and more extreme and that cold spells have become shorter and milder. For example, the western Europe heat wave of 2003 was responsible for upwards of 70,000 deaths and was the warmest summer there in more than 600 years (Robine et al., 2008). No single event like that can be reliably attributed to climate change, but it is consistent with expectations for the future. Within the United States, hot days, hot nights, and heat waves have become more frequent in recent decades and were the leading cause of weather-related morbidity and mortality during 1970–2004 (USGCRP, 2009).

On an urban scale, the heat-island effect contributes to local temperature increase. For example, the urban heat island around Phoenix, Arizona, raises minimum nighttime temperatures by as much as 12.6°F (7°C) (Brazel et al., 2000). When increased ozone events occur simultaneously with heat waves, mortality can rise by 175% (Filleul, 2006). As extremely hot days tend to be associated with high pressure and stagnant air-circulation patterns, ground-level ozone, $PM_{2.5}$, particulate sulfate, and organic carbon have been found to correlate strongly in summer months (NRC, 2008).

Measurements of rainfall indicate that moist regions of the globe are getting wetter and semiarid regions are becoming drier; this is consistent with an intensification of the hydrologic cycle. In situ and space-based precipitation observations indicate that both global precipitation and extreme rainfall events are increasing. Total runoff is increasing but shows substantial regional variability (cf. USGCRP, 2009). In the United States, the amount of precipitation falling in the heaviest 1% of rain events increased by 20% in the past century, and total precipitation by 7%. Over the past century, there was a 50% increase in the frequency of days with precipitation of more than 10 cm in the upper Midwest. Heavy rains can lead not only to flooding but to a greater incidence of sewage overflows, contaminated drinking water, and waterborne diseases, such as cryptosporidiosis and giardiasis. Rivers and lakes are freezing later and thawing earlier with serious implications for flooding. The manner in which increased temperature and decreased rainfall covary in the western United States has led to a 400% increase in western wildfires in recent decades (Westerling et al., 2006). Drought and possible changes in irrigation practices could induce more frequent windblown-dust storms, which constitutes an air-quality effect with potential public-health consequences.

Literature Regarding Projected Climate Change

Observations like those summarized above needed to be supplemented with models that project potential conditions. Such predictions are essential for guiding policy because of the long lag times associated with changes in our built environments. Policy-makers need to be able to anticipate future change before it occurs to be able to plan appropriately.

Projections of climate change are derived from the output of numerical models similar to the models used for numerical weather prediction albeit at coarser resolution. For day-to-day weather prediction, with a spatial resolution of tens of kilometers, the prediction is influenced by the initial conditions and the observed state of the atmosphere. In contrast, a climate projection of the general state of the atmosphere—global mean temperature over the next 100 years—is influenced by changes in the concentration of heat-trapping greenhouse gases and coupling of the atmosphere to the ocean, land surface, and cryosphere.

At the time of the first Intergovernmental Panel on Climate Change (IPCC) assessment report in 1990, the best resolution of climate models was around 500 km; for the fourth IPCC assessment report (AR4) in 2007, the best resolution was around 100 km; and to support the fifth IPCC assessment, due in 2013, some climate-change models are being run at resolutions of tens of kilometers. The importance of greater and greater resolution means that future IPCC assessments will move away from global mean metrics of climate change (such as temperature and sea-level rises) and toward a much greater emphasis on the anticipated changes at regional levels. As with spatial resolution, the climate projections run since 1990 have focused on the mean states of future climate for, say, a decade in the future, that is, 2089–2099. Because extreme climatic events often take place at the regional level on relatively short time scales, time and space become coupled. Hence, to simulate the change in extreme or high-intensity climate events, such as storms or floods, high resolution in climate models is a necessity, but it has been limited in the past by the capability of high-performance computing platforms. It must be remembered, though, that the usefulness of high-resolution models is limited by uncertainties in information supplied by the larger-scale models they depend on and the natural variability in the climate (USCCSP, 2008).

The findings of the fourth IPCC assessment (2007) indicate that global average surface temperatures are projected to rise from the 1980–1999 average by 1.1–6.4°C by the end of the 21st century. Global sea level will rise by 0.8–2 m by 2100. The effects of global sea-level rise will be exacerbated at the regional level along the eastern seaboard of the United States by a likely increase in the intensity of Atlantic hurricanes and resulting storm surge. Heat waves will become more intense, more frequent, and longer-lasting, and the frequency of cold extremes will continue to decrease. By 2100, the number of heat-wave days is expected to double in Los Angeles and quadruple in Chicago (USGCRP, 2009). The intensity of precipitation events is also expected to continue to increase and to result in more frequent heavy downpours and floods, most notably in wetter regions, and droughts are expected to become more common in semiarid regions. That projected acceleration of the hydrological cycle suggests that rainfall will become more concentrated into intense events with longer, hotter dry periods between them. Implications for the continental United States are that the northern tier of states will become wetter with attendant increased runoff and that the southern states will become drier, especially in the West. In the face of those changing patterns of temperature, precipitation, and extreme events, the range and effects of pathogens and pests are also expected to change.[1]

Beyond anecdotal evidence and extrapolation, there has been little study of how climate change will influence the indoor environment from the perspective of adverse effects on human health. Given that climate-change projections with regional specificity are only now becoming available, that may not be surprising. However, the advent of climate-change projections on regional scales makes a number of types of research possible.

In the future, the climate-modeling community will strive for higher and higher resolution of climate models by increasing the resolution of global models everywhere and by using the output of current global models as input into regional and urban models with downscaling techniques. The move from climate models to so-called *Earth System Models*—in which aspects of chemistry, biology, and ecosystem functioning are incorporated at the junction of the physical climate system and biogeochemical cycling—represents the next grand challenge to the climate-science community (NRC, 2010b).

ADVERSE EXPOSURES ASSOCIATED WITH CLIMATE-CHANGE–INDUCED ALTERATIONS IN THE INDOOR ENVIRONMENT

Indoor environmental conditions exert considerable influence on health (ASHRAE, 2010; HHS, 2005, 2010), learning (NRC, 2006), and productivity (Fisk and Rosenfeld, 1997; Mendell and Heath, 2005; NRC, 2006; Seppänen and Fisk, 2004). Fisk and Rosenfeld (1997) estimated that poor environmental conditions and indoor contaminants cost the US economy tens of billions of dollars a year in exacerbation of illnesses, allergenic symptoms that include asthma, and lost productivity. Research conducted by the US Environmental Protection Agency suggests that such indoor contaminants as radon, secondhand smoke, and volatile organic compounds contribute to tens of thousands of excess deaths a year, with premature deaths from pollutants emitted indoors equivalent to the impact of outdoor particulate pollution (Mudarri, 2010). Reviews of the scientific literature by Institute of Medicine committees (2000, 2004) concluded that there

[1] This topic is addressed in Chapter 6.

was evidence of an association between new-onset asthma and indoor dampness, molds, and dust mites. The 2006 National Research Council report *Green Schools: Attributes for Health and Learning* concluded that moisture problems, inadequate ventilation, and airborne contaminants in public schools contribute to suboptimal learning and absenteeism among teachers, administrators, and students.

Indoor environmental quality is a function of four general factors: macroenvironment, building infrastructure, occupant furnishings and activities, and occupant health and perceptions. These factors are detailed below.

Macroenvironment factors include such items as outdoor pollution, climate and weather conditions, and soil conditions, including geologic features that affect the risk of radon emission. With reference to climate change, the confluence of extreme precipitation events, impermeable surfaces, and soil conditions influences the effect of water on structures. How water is managed around buildings and the integrity of a structure will help to determine moisture transport and its effects on indoor environments.

Building infrastructure and building component systems have both direct and indirect influences on indoor contaminants. Indoor environmental quality is a function of the interrelationships of a building's foundations; floors, walls, and roofs; heating, ventilation, and air-conditioning (HVAC) systems; electric and plumbing systems; materials; and furnishings. The building envelope's tightness or porosity; the integrity of foundations, roofs, and windows; and other planned and unplanned openings all influence the infiltration of outdoor moisture and air pollutants. Studies estimate that about half the outside air that enters even a mechanically ventilated building finds its way in through unducted pathways (Persily, 1997).

Building ventilation systems provide conditioned air and dilute internally generated contaminants. HVAC systems, for example, affect a variety of indoor environmental factors, including pollutant levels, temperature, humidity, noise, air quality, moisture control, and odors. The location of air intakes, the efficiency of ventilation filters, and operating practices all affect the amount and quality of outdoor air used to ventilate indoor spaces. The optimum size and capacity of an HVAC system depend on the orientation of the building, the total floor area, the quality of insulation, the number of windows, and other factors. Other components, such as plumbing and electric systems, often create penetrations between floors that contribute to unplanned pathways for contaminant movement.

There are numerous other examples of interrelationships between the design and operation of a building system and its indoor environmental quality. Generally speaking, indoor environmental quality deteriorates if buildings are not properly designed, systems are not operated appropriately, or needed maintenance and repairs are not performed or are deferred (NRC, 2006).

Structural features (foundations, façade, thermal bridges, roof design, and the like), details of construction specifications, and integrity of construction can also influence indoor conditions. Those elements affect the bulk, capillary and vapor transport of water, and passive or active movement of air through the structure.

Occupant furnishings and activities play a central role in influencing indoor conditions, initially through design and specifications of building systems and materials. Occupants, owners, facility managers, purchasing agents, interior designers, and others make many decisions about furnishings, decorative materials, cleaning products, appliances, and equipment that can emit particles and gases into the interior of buildings. Occupants make myriad choices related to product use, maintenance of products, equipment, and appliances and undertake actions that influence ventilation and hence contaminant concentrations and moisture. "Sick-building" investigations have shown indoor problems related to materials' off-gassing (of formaldehyde, for example) that, in some cases, was precipitated or aggravated by other factors related to design, operation and use, or maintenance (Oliver and Shackleton, 1998; Šeduikytė and Bliūdžius, 2003; Seppänen and Fisk, 2004).

Occupant health and perceptions, which influence susceptibility and response to contaminant exposures and indoor conditions, are perhaps the most complicated component of indoor environmental quality because of the inherent variability in human expectations and vulnerabilities. The variability makes it difficult to draw inferences from scientific research for codification in ventilation, comfort, material performance, and health standards in the many different types of indoor environments.

Climate Change Concerns for Indoor Environments and Possible Health Risk

This report examines the influences that changing weather patterns and shifting climate regimes may have on factors that affect indoor environments and the health of occupants. Figure 2-1 illustrates how climate-change–induced scenarios could affect building operations and indoor environments and possibly lead to human health effects through exposures to physical, chemical, and biologic stresses. Several of the scenarios involve moisture intrusion into buildings directly or as a result of condensation. Prolonged heat waves will heat the thermal mass of structures to the extent that the radiant-heating component will become more important indoors. Warmer ambient environments will mean more air-conditioning use in buildings, which in turn alters ventilation and dew points within structures. Climate change models project increases in hydrocarbon emissions and

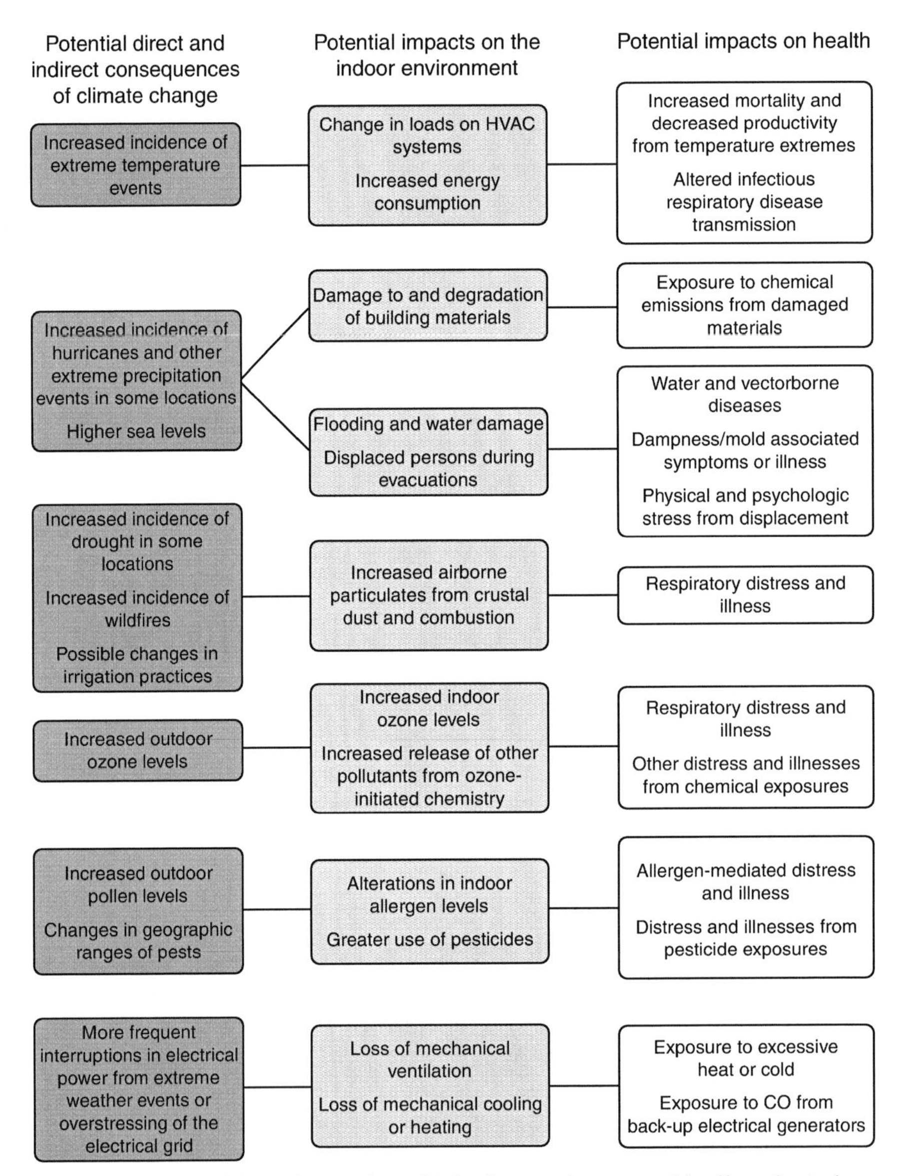

FIGURE 2-1 Possible pathways by which climate change could affect the indoor environment and health (adapted from Su, undated).

concomitant increases in outdoor ozone concentrations. They, in turn, have implications for ozone penetration indoors and later chemical reactions.

The committee organized its examination of the literature regarding potential alterations in indoor environmental quality induced by climate change into five primary categories: air quality; dampness, moisture and flooding; infectious agents and pests; thermal stress; and building design, construction, operation, maintenance, and retrofitting. The divisions are in some respects arbitrary—for example, damp spaces provide a hospitable environment for some pests and infectious agents and thus affect air quality—but they are a means of rationalizing a complex set of circumstances that influence the health of building occupants. Chapters 4–8 address the science regarding them.

TIME SPENT IN THE INDOOR ENVIRONMENT

Exposure is a function of pollutant levels and the time spent in contact with the pollutants. Several studies have examined where people spend their time, how long they are in those environments, and, in some cases, the extent of their physical activity in the environments. An understanding of the amount of time that people spend indoors and the variations in different segments of the population is central to the evaluation of the risks associated with potential alterations in indoor environmental quality induced by climate change. Information on time spent in particular environments is also relevant to developing strategies to reduce problematic exposures and in turn to improve health.

The majority of people's time in the United States is spent indoors, whether in residences, in schools, or in workplaces. According to the 1994 National Human Activity Pattern Survey, the average person spends just over 92% of his or her time indoors; of that time indoors, almost 70% is spent in one's residence (Klepeis et al., 2001). Care must be exercised in generalizing from that, inasmuch as some studies include time spent in vehicles—typically 4–6% of the day—in accounting for indoor time (Dales et al., 2008; Klepeis et al., 2001; Leech, 2002; Zhang and Batterman, 2009).

Researchers have also examined the time spent indoors in other countries. In a 1998 study in Italy, it was found that people spent 84% of their time indoors, with 64% of that time at home and 3.4% in vehicles (Simoni et al., 1998). Another study in different cities representing the seven regions of Europe found that people spent 90% indoors—58% at home, 25% at work, and 7% in vehicles and other indoor public environments (Schweizer, 2007). Studies in Canada found that about 89–90% of time is spent indoors (Kim et al., 2005; Wu et al., 2007). Even more striking, those in New Zealand tend to spend ~94% indoors, 5% of it in transit (Baker et al., 2007).

When different regions and times of the year were looked at, few differences were noted in how the average adult spent his or her time. For time spent indoors in residential environments, no significant difference was found between the northeastern, midwestern, southern, and western regions of the United States (EPA, 1996; Klepeis et al., 2001). On the average, people were in their homes 69.4–70.7% of the time (EPA, 1996; Klepeis et al., 2001). Similarly, the time of the year only showed a small difference: 67.9% of the time was spent indoors during spring and 71.9% in winter (EPA, 1996). The one variation was between weekdays and weekends: the mean time spent in residences during weekdays and weekends was 67.1% and 74.6%, respectively (EPA, 1996).

It appears that adults living across all US Census regions tend to spend about 6% of their day in vehicles, with little contrast between the seasons (EPA, 1996). In contrast with time spent in residences during the weekdays and weekends, there was no difference in time spent in vehicles (EPA, 1996).

Children, particularly young children, spend a large fraction of their time indoors. Children under 2 years old tend to spend the most time inside, just under 94% (Cohen-Hubal et al., 2000; EPA, 2009). Time spent indoors continued to be 83–94% throughout childhood, including 19% in school (EPA, 1996, 2009). Younger children tended to spend more of their time at home than older children but only during the traditional school year (Silvers, 1994). It is necessary to note that older children are not necessarily spending more time outdoors when they are not at home; in fact, they often are spending more time in the school environment (Silvers, 1994). During summer, younger children were more apt not to spend time at home and older children more apt to spend time at home (Silvers, 1994).

There has been a trend toward students' spending less time in school and participating less in sports and other outdoor activities than 30 years ago (Juster et al., 2004). In 1981, children spent about 75 min/day outdoors (Juster et al., 2004) while in 2003, they spent only 50 min (Juster et al., 2004). That shift is peculiar to children: time spent indoors not only has increased slightly but has shifted between time spent in the residence and time spent in other indoor facilities. In adults, however, time spent indoors has remained constant over the past several decades (Klepeis, 2001).

A cohort study performed in New York, New Jersey, Pennsylvania, Washington, Oregon, and California looked at seasonal differences. It found that children 5–12 years old increased their time spent indoors only in summer (Silvers et al., 1994). One interesting point is that that did not vary from one region to another (Klepeis et al., 2001; Silvers et al., 1994).

The elderly tend to spend more time indoors, particularly in their residences, than do their younger counterparts (Berry, 1991; Franklin, 2004; Geller and Zenick, 2005; Kenney and Munce, 2003; Klinenberg, 2002).

TABLE 2-1 Percentage of Time Spent Indoors as a Function of Age

Population, age in years	Fraction of Time Spent in Residence, %	Fraction of Total Time Spent Indoors
General population[a]	69	86.5–91.6
Children and youth[b]		
Birth to <1	75.7	94
1 – <2	72.7	94
2 – <3	67.3	91.4
3 – <6	66.0	88.8
6 – <11	60.6	83.4
11 – <16	60.8	87.5
16 – <21	56.9	86.6
Elderly (>64)[c]	81.6–95	

[a] Bernstein (2008), Dales (2008), Klepeis (2001).
[b] EPA (2009).
[c] Berry (1991), EPA (1996).

Table 2-1 summarizes information on time spent indoors in the United States as a function of age.

Some researchers have suggested that shifts in ambient conditions due to climate change will lead to people spending more time indoors (Bluyssen, 2009; Samet, 2009). This is plausible, given that sheltering indoors is a common response to extreme weather conditions such as high heat. However, the lack of regional differences in time spent indoors in the United States suggests that adaptation also plays a role in this decision and insufficient information exists to draw confident conclusions about whether and how such factors will influence future behavior.

CLIMATE CHANGE AND VULNERABLE POPULATIONS

Segments of the population will vary in their ability to adapt to climate change–induced alterations in the indoor environment, depending on their circumstances. This section addresses a number of factors that might influence whether particular populations are more vulnerable to adverse effects.

Vulnerability relates to the balance between susceptibility factors and factors that increase the resilience of populations to environmental stressors (Balbus and Malina, 2009). It is a dynamic characteristic and can include the geographic region in which one resides and the adaptive capacity of an individual, including the presence of chronic medical conditions, low socioeconomic conditions, infancy or old age, and living in an isolated or segregated area (Shonkoff et al., 2009). Racial and ethnic minorities may be at great risk for health conditions related to climate change.

Kelly and Adger (2000) write that a person's vulnerability is determined by access to resources and the diversity of income sources and by social status of the person or the person's household in the community. The ability of a person to adapt is influenced by intrinsic factors (such as age and health) and extrinsic factors (such as housing and the availability of and ability to go to shelters during extreme weather events). Poverty is therefore an important indicator of individual vulnerability to climate change and is related to marginalization and lack of resources. Poverty affects vulnerability through people's expectations of the effects of hazards and their ability to marshal resources to alleviate risks.

Chapter 7 addresses the literature on the role that biologic vulnerability and economic and social circumstances play in determining the risk of health effects of exposure to heat. Several other aspects of climate change's effects on the indoor environment might also affect various segments of the population disproportionately.

Susceptibility to such changes as increased incidence of extreme weather events, high humidity, and expanded ranges of some pests can be expected to be influenced by physiologic factors. Biologic sensitivity may be related to a person's developmental stage, pre-existing chronic medical conditions, acquired factors (such as immunity), and genetic factors (such as metabolic enzyme subtypes that play a role in sensitivity to toxic substances) (Balbus and Malina, 2009). Children have been shown to be more vulnerable to the effects of exposure to a number of indoor chemicals as a result of their metabolic rates, body size, behaviors, immature immune responses, and still-developing ability to detoxify substances (Faustman et al., 2000). Human and experimental studies show that the fetus and infant are more sensitive than adults to many environmental toxicants, including residential pesticides (Perera et al., 2005; WHO, 1986; Whyatt et al., 2004). In addition, some types of medications—including antipsychotic, antiparkinsonian, and anticholinergic drugs—may increase vulnerability to environmental insults (Brown and Walker, 2008; Kenny et al., 2010; Luber and McGeehin, 2008; WHO, 2004).

Increasing temperatures and increasing humidity associated with climate change are expected to result in changing patterns of insects and rodents. People in multifamily urban dwellings where pesticides are commonly used may be at increased risk for exposure and have little control over the pesticides that might be used in their buildings. Children and pregnant women will be most vulnerable to the health consequences of pesticide exposure. Pesticides sprayed outdoors can also find their way indoors through air exchange or be brought in on clothing, on skin, and especially on shoes. People living close to agricultural operations may also be at particularly high risk.

Homes in low-income areas tend to have greater occupant density, which exposes more people to pollutants in the indoor environment. Although one study found that type of building material did not increase vulnerability to climate change (Kovats and Hajat, 2008), other research indicates that brick homes that have high thermal mass and top-floor apartments that have poor ventilation and closed windows are associated with increased mortality during heat waves relative to other buildings (Mirchandani, 1996; Vandentorren et al.; 2006). Older homes that have poor insulation and poor ventilation may expose families to increased risk from water events, rising humidity, and pest infestation. Low-income homes tend to be older, and this could be associated with "leakier" home environments and more contamination with outside pollutants (Chan et al., 2005). Leakier homes could also be at greater risk for water damage and infestation with rodents or insects. Some studies have found that multi-family units are not necessarily "leakier" but instead have lower rates of ventilation, which could increase the risk of health effects of exposure to indoor sources of pollutants (Zota et al., 2005).

Home ownership also has an influence on occupant health, with home owners reporting better health status and better health outcomes than renters (Kuh et al., 2002; Pollack et al., 2010; Robert and House, 1996; Wadsworth et al., 1999). There are several potential reasons for this; most centered on differences in wealth and socioeconomic status.

Renting may leave less disposable income for health care. Renters tend to have lower incomes than homeowners, and a larger percentage of renters' incomes tend to be allocated to rent than homeowners' incomes are allocated to mortgages. Minorities and those with lower incomes are more likely to rent than own, and those who had difficulty paying rent and utility bills were less likely to seek out medical care when needed (Kushel et al., 2006).

Neighborhoods with high levels of homeownership tend to be neighborhoods with higher wealth and socioeconomic status, thus also influencing the physical condition of the housing unit based on neighborhood conditions (Kearns et al., 2000; RWJ, 2008). The American Housing Survey found that older tenants tend to live in more expensive, yet lower quality housing than their home-owning counterparts (Muller et al., 2001).

Additionally, renters are at a disadvantage in that they have less control over modifications made in their residences. If the owners delay or ignore requests for improvements to the housing unit, the tenants are left with little recourse (Pynoos and Nishita, 2003).

Homeowners are also more likely to take precautionary measures against possible health hazards, which may be in part due to longer length of time spent residing in the same housing unit and the larger financial investment placed in the home. This is seen most prominently in the case of radon. The National Health Interview Surveys in 1994 and 1998 found that radon awareness and testing differed between homeowners and renters (Larsson et al., 2009). Those who owned a single family home or townhome were more likely to have heard of radon and to get their homes tested than those who rented apartments or condos (Larsson et al., 2009). And, a survey of New York residents found that homeowners were more likely to perform radon mitigation actions than renters (Wang et al., 1999).

Having homeowner's or renter's insurance allows families to adapt to events associated with climate change. Insured families whose homes experience water damage can obtain repairs quickly whereas uninsured families are forced to vacate their homes or live in substandard, damp environments for long periods. In general, the poorest households are most likely to have the poorest air quality, whether because of a lack of air conditioning or because they are underinsured with respect to repairing damage from climate-change–related moisture (Fothergill and Peek, 2004; Thomalla et al., 2006).

Studies indicate that there are regional differences in health outcomes (Fisher et al., 2009; Halverson et al., 2004). Balbus and Malina (2009), who focused their analysis on potentially vulnerable populations for climate change health effects, assert that populations in certain parts of the United States may experience "increased risks for specific climate-sensitive health outcomes" and that "[s]ome regions' populations may in fact experience multiple climate-sensitive health problems simultaneously." The researchers offered four examples—locations of past hurricane landfalls, past extreme heat events, high concentrations of population 65 years of age or older, and cases of West Nile virus—to illustrate how geographic, demographic, and climatic factors might influence regional vulnerabilities. The 2010 National Academies report *Adapting to the Impacts of Climate Change* (NRC, 2010a) summarized potential regional climate-change impacts in a table that is excerpted below (Table 2-2). Such projections must be viewed with great caution, though. Among the uncertainties listed by the NRC report is "an inability to attribute explicitly many observed changes at local and regional scales to climate change."

In summary, vulnerability to health effects associated with climate change and indoor environmental quality will depend on the process under scrutiny and will be the result of an interaction of extrinsic and intrinsic factors. Most of the adaptation to climate change and resulting indoor environmental quality will depend on changes implemented by the residents of homes. Some populations that lack the resources to change their homes' ventilation systems or to repair water damage will suffer from increasing indoor temperatures and increasing humidity. Poorer communities, including people who live in developing countries, will be very susceptible to health effects of climate change and the indoor environment. Children, the elderly, and people who have chronic health conditions will be most susceptible to the effects of poor indoor environmental quality, and people who have

TABLE 2-2 Summary of Selected Potential Regional Climate-Change–Related Impacts

United States Census Region	Climate Related Impacts						
	Degraded Air Quality	Urban Heat Islands	Wildfires	Heat Waves	Tropical Storms	Extreme Rainfall with Flooding	Sea Level Rise
New England ME VT NH MA RI CT	●	●		●		●	●
Middle Atlantic NY PA NJ DE MD	●	●		●	●	●	●
East North Central WI MI IL IN OH	●	●		●		●	
West North Central ND MN SD IA NE KS MO		●		●		●	
South Atlantic WV VA NC SC GA FL DC	●	●	●	●	●	●	●
East South Central KY TN MS AL		●		●	●	●	●
West South Central TX OK AR LA	●	●	●	●	●	●	●
Mountain MT ID WY NV UT CO AZ NM	●	●	●	●		●	
Pacific AK CA WA OR HI	●	●	●	●	●	●	●

(excerpted from NRC, 2010a; adapted from CCSP, 2008)

pre-existing allergic conditions or respiratory diseases may find that their conditions are worsened.

CONCLUSIONS

On the basis of its review of the papers, reports, and other information presented in this chapter, the committee has reached the following conclusions related to the potential effects of climate change on the indoor environment and health, to time spent in indoor environments, and to vulnerability. Later chapters revisit some of these issues in greater detail and offer additional observations.

- The frequency and intensity of some extreme weather events, such as heavy precipitation and heat waves, are increasing. Models suggest that there will be important regional differences in these events: some areas of the country will become drier and others and wetter.
- There is a lack of understanding of the linkages between climate change, indoor environmental quality, and health.
- Because people spend the vast majority of their time in indoor environments, they will encounter many of the effects of climate change indoors.
- Vulnerable populations will be disproportionately affected by climate change and its adverse effects on indoor environmental quality. Vulnerable populations include those who have less economic ability to adapt to or mitigate the effects of changes in their indoor environment and those whose age or health status renders them more susceptible to environmental stresses or insults.

REFERENCES

ASHRAE (American Society of Heating, Refrigeration, and Air-Conditioning Engineers). 2010. *Ventilation for acceptable indoor air quality (Standard 62.1)*. Atlanta, GA: ASHRAE.

Baker M, Keall M, Au E, Howden-Chapman P. 2007. Home is where the heart is—most of the time. *Journal of the New Zealand Medical Association* 120(1264):U2769.

Balbus JM, Malina C. 2009. Identifying vulnerable subpopulations for climate change health effects in the United States. *Journal of Occupational and Environmental Medicine* 51(1): 33-37.

Bernstein J, Alexis N, Bacchus H, Bernstein I, Fritz P, Horner E, Li N, Mason S, Nel A, Oullette J, Reijula K, Reponen T, Seltzer J, Smith A, Tarlo SM. 2008. The health effects of non-industrial indoor air pollution. *Journal of Allergy and Clinical Immunology* 121(3):585-591.

Berry M. 1991. Indoor air quality: Assessing health impacts and risks. *Toxicology and Industrial Health* 7(5-6):179-186.

Bluyssen PM. 2009. *The indoor environment handbook: How to make buildings healthy and comfortable.* London, UK: Earthscan Publications Ltd. (Taylor & Francis Group).

Brazel AN, Selove RV, Heisler G. 2000. The tale of two climates—Baltimore and Phoenix urban LTER sites. *Climate Research* 15(2):123-135.

Brown S, Walker G. 2008. Understanding heat wave vulnerability in nursing and residential homes. *Building Research & Information* 36(4):363-372.

CCSP (Climate Change Science Program). 2008. *Analyses of the effects of global change on human health and welfare and human systems. Synthesis and assessment Product 4.6.* Report by the US Climate Change Science Program and the Subcommittee on Global Change Research, KL Ebi, FG Sussman, and TJ Wilbanks (authors), edited by JL Gamble. Washington, DC: US Environmental Protection Agency.

Chan W, Nazaroff WW, Price PN, Sohn MD, Gadgil AJ. 2005. Analyzing a database of residential air leakage in the United States. *Atmospheric Environment* 29:3445-3455.

Cohen-Hubal E, Sheldon L, Burke J, McCurdy, T, Berry M, Rigas M, Zartarian V, Freeman N. 2000. Children's exposure assessment: A review of factors influencing children's exposure, and the data available to characterize and assess that exposure. *Environmental Health Exposure* 108(6):475-486.

Dales R, Ling L, Wheeler A, Gilbert N. 2008. Quality of indoor residential air and health. *Canadian Medical Association Journal* 179(2):147-152.

EPA (Environmental Protection Agency). 1996. *Descriptive statistics tables from a detailed analysis of the National Human Activity Pattern Survey (NHAPS) Data.* Washington, DC: EPA/600/R-96-148.

EPA. 2009. *Child-specific exposure factors handbook 2008.* Washington, DC: EPA/600/R-06/096F.

Faustman EM, Silbernagel SM, Fenske RA, Burbacher TM, Ponce RA. 2000. Mechanisms underlying children's susceptibility to environmental toxicants. *Environmental Health Perspectives* 108(Suppl 1):13-21.

Filleul L, Cassadou S, Medina S, Fabres P, Lefranc A, Eilstein D, Le Tertre A, Pascal L, Chardon B, Blanchard M, Declercq C, Jusot JF, Prouvost H, Ledrans M. 2006. The relation between temperature, ozone, and mortality in nine French cities during the heat wave of 2003. *Environmental Health Perspectives* 114(9):1344-1347.

Fisher ES, Bynum JP, Skinner JS. 2009. Slowing the growth of health care costs—lessons from regional variation. *New England Journal of Medicine* 360(9):849-852.

Fisk WJ, Rosenfeld AH. 1997. Estimates of improved productivity and health from better indoor environments. *Indoor Air* 7:158-172.

Fothergill A, Peek L. 2004. Poverty and disasters in the United States: A review of recent sociological findings. *Natural Hazards Journal* 32(1):89-110.

Franklin C. 2004. Lessons from a heat wave. *Intensive Care Medicine* 30(1):167.

Geller A, Zenick H. 2005. Aging and the environment: A research framework. *Environmental Health Perspectives* 113(9):1257-1262.

Halverson JA, Ma L, Harner EJ 2004. *An analysis of disparities in health status and access to health care in the Appalachian region.* Washington, DC: The Appalachian Regional Commission.

HHS (US Department of Health and Human Services). 2005. *Workshop on healthy indoor environments. A report of the surgeon general.* Rockville, MD: Office of the Surgeon General.

HHS. 2010. *How tobacco smoke causes disease: The biology and behavioral basis for smoking-attributable disease.* A Report of the Surgeon General. Rockville, MD: Office of the Surgeon General.

IPCC (Intergovernmental Panel on Climate Change). 2007. *Climate change, 2007: The physical science basis, Contribution of Working Group I to the fourth assessment report of the Intergovernmental Panel on Climate Change*. Cambridge: Cambridge University Press.

Juster F, Ono H, Stafford F. 2004. Changing times of American youth: 1981-2003. Child Development Supplement.

Kearns A, Hiscock R, Ellaway A, Macintyre S. 2000. "Beyond four walls": The psychosocial benefits of home. Evidence from West Central Scotland. *Housing Studies* 15(3):387-410.

Kelly PM, Adger WN. 2000. Theory and practice in assessing vulnerability to climate change and facilitating adaptation. *Climatic Change* 47:325-352.

Kenney W, Munce T. 2003. Invited review: Aging and human temperature regulation. *Journal of Applied Physiology* 95(6):2598-2603.

Kenny GP, Yardley J, Brown C, Sigal RJ, Jay O. 2010. Heat stress in older individuals and patients with common chronic diseases. *CMAJ: Canadian Medical Association Journal* 182(10):1053-1060.

Kim D, Sass-Kortsak A, Purdham J, Dales R, Brook J. 2005. Sources of personal exposure to fine particles in Toronto, Ontario, Canada. *Journal of Air & Waste Management* 55(8):1134-1146.

Klepeis NE, Nelson WC, Ott WR, Robinson JP, Tsang AM, Switzer P, Behar JV, Hern SC, Engelmann WH. 2001. The National Human Activity Pattern Survey (NHAPS): A resource for assessing exposure to environmental pollutants. *Journal of Exposure Analysis and Environmental Epidemiology* 11:231-252.

Klinenberg E. 2002. *Heat wave: A social autopsy of disaster in Chicago*. Chicago: University of Chicago Press.

Kovats RS, Hajat S. 2008. Heat stress and public health: A critical review. *Annual Review of Public Health* 29:41-55.

Kuh D, Hardy R, Langenber C, Richards M, Wadsworth M. 2002. Mortality in adults aged 26-54 years related to socioeconomic conditions in childhood and adulthood: Post war birth cohort study. *British Medical Journal* 325(7372):1076-1080.

Kushel MB, Gupta R, Gee L, et al. 2006. Housing instability and food insecurity as barriers to health care among low-income Americans. *Journal of General Internal Medicine* 21(1):71-77.

Larsson LS, Hill WG, Odom-Maryon T, Yu P. 2009. Householder status and residence type as correlates of radon awareness and testing behaviors. *Public Health Nursing* 26(5): 387-395.

Leech J. 2002. Outdoor air pollution epidemiologic studies. *American Journal of Respiration and Critical Care Medicine* 161(3):A308.

Luber G, McGeehin M. 2008. Climate change and extreme heat events. *American Journal of Preventative Medicine* 35(5):429-435.

Mendell MJ, Heath GA. 2005. Do indoor pollutants and thermal conditions in schools influence student performance? A critical review of the literature. *Indoor Air* 15:27-52.

Mirchandani HG, McDonald G, Hood IC, Fonseca C. 1996. Heat-related deaths in Philadelphia—1993. *American Journal of Forensic Medicine and Pathology* 17:106-108.

Mudarri, D. 2010. *Public health consequences and cost of climate change impacts on indoor environments*. Washington, DC: EPA.

Muller C, Gnanasekaren K, Knapp K, Dushi I. 2001. *Older homeowners and renters in six US cities: Housing and economic resources*. International Longevity Center-USA. New York: New York.

NRC (National Research Council). 2006. *Green schools: Attributes for health and learning*. Washington, DC: The National Academies Press.

NRC. 2008. *Estimating mortality risk reduction and economic benefits from controlling ozone air pollution*. Washington, DC: The National Academies Press.

NRC. 2010a. *Adapting to the impacts of climate change*. Washington, DC: The National Academies Press.
NRC. 2010b. *Advancing the science of climate change*. Washington, DC: The National Academy Press.
NRC. 2010c. *Informing an effective response to climate change*. Washington, DC: The National Academies Press.
NRC. 2010d: *Limiting the magnitude of future climate change*. Washington, DC: The National Academies Press.
Oliver LC, Shackleton BW. 1998. The indoor air we breathe. *Public Health Reports* 113(5): 398-409.
Perera FP, Rauh V, Whyatt RM, Tang D, Tsai WY, Bernert JT, Tu YH, Andrews H, Barr DB, Camann DE, Diaz D, Dietrich J, Reyes A, Kinney PL. 2005. A summary of recent findings on birth outcomes and developmental effects of prenatal ETS, PAH, and pesticide exposures. *Neurotoxicology* 26(4):573-587.
Persily A. 1997. Evaluating building IAQ and ventilation with indoor carbon dioxide. *ASHRAE Transactions* 102(2).
Pollack C, Egerter S, Sadegh-Nobari T, Dekker M. 2010. *Issue brief 2: Housing and health*. Robert Wood Johnson Foundation Commission to Build a Healthier America.
Pollack CE, Griffin BA, Lynch J. 2010. Housing affordability and health among homeowners and renters. *American Journal of Preventive Medicine* 39(6):515-521.
Pynoos J, Nishita C. 2003. The cost and financing of home modifications in the United States. *Journal of Disability Policy Studies* 14(2):68-73.
Robert S, House J. 1996. SES differentials in health by age and alternative indicators of SES. *Journal of Aging and Health* 8(3):359-388.
Robine JM, Cheung SLK, Le Roy S, Oyen HV, Griffiths C, Michel JP, Herrmann FR. 2008. Death toll exceeded 70,000 in Europe during the summer of 2003. *Comptes Rendus Biologies* 331(2):171-175.
RWJ (Robert Wood Johnson Foundation Commission to Build a Healthier America). 2008. *Issue brief 2: Housing and health*.
Samet JM. 2009. *Adapting to climate change: Public health*. Washington, DC: Resources for the Future. http://www.rff.org/rff/documents/RFF-Rpt-Adaptation-Samet.pdf.
Schweizer C, Edwards RD, Bayer-Oglesby L, Gauderman WJ, Ilacqua V, Jantunen MJ, Lai HK, Nieuwenhuijsen M, Künzli N. 2007. Indoor time—microenvironment—activity patterns in seven regions of Europe. *Journal of Exposure Science and Environmental Epidemiology* 17(2):170-181.
Šeduikytė L, Bliūdžius R. 2003. Indoor air quality management. *Environmental Research, Engineering & Management* 1(23):21-30.
Seppänen OA, Fisk WJ. 2004. Summary of human responses to ventilation. *Indoor Air* 14(Suppl 7):102-118.
Shonkoff et al. 2009. *Environmental health and equity impacts from climate change and mitigation policies in California: A review of the literature*. California Environmental Protection Agency, Air Resources Board. CEC-500-2009-038-D.
Silvers A. 1994. How children spend their time: A sample survey for use in exposure and risk assessments. *Risk Analysis* 14(6):931-944.
Simoni M, Biavati P, Carrozzi L, Veigi G, Paoletti P, Matteucci G, Ziliani GL, Ioannilli E, Sapigni T. 1998. The Po River Delta (North Italy) indoor epidemiological study: Home characteristics, indoor pollutants, and subjects' daily activity pattern. *Indoor Air* 8(2):70-79.
Su JH. Undated. *The implication of weather extremes and a rapidly changing climate to indoor environments*. PowerPoint presentation provided by the author.

Thomalla F, Downing T, Spanger-Siegfried E, Han G, Rockström J. 2006. Reducing hazard vulnerability: Towards a common approach between disaster risk reduction and climate adaptation. *Disasters* 30(1):39-48.

USCCSP (US Climate Change Science Program). 2008. *Climate models: An assessment of strengths and limitations*. Bader DC, Covey C, Gutowski WJ, Held IM, Kunkel KE, Miller RL, Tokmakian RT, Zhang MH, primary authors. Washington, DC: Department of Energy, Office of Biological and Environmental Research.

USGCRP (US Global Change Research Program). 2009. *Global climate change impacts in the United States*. New York: Cambridge University Press.

Vandentorren S, Bretin P, Zeghnoun A, Mandereau-Bruno L, Croisier A, Cochet C, Ribéron J, Siberan I, Declercq B, Ledrans M. 2006. August 2003 heat wave in France: Risk factors for death of elderly people living at home. *European Journal of Public Health* 16:583-591.

Wadsworth M, Montgomery S, Bartley M. 1999. The persisting effect of unemployment on health and social well-being in men early in working life. *Social Science and Medicine* 48(10):1491-1499.

Wang Y, Ju C, Stark AD, Teresi N. 1999. Radon mitigation survey among New York State residents living in high radon homes. *Health Physics* 77(4):403-409.

Westerling AL, Hidalgo H, Cayan DR, Swetnam T. 2006. Warming and earlier spring increases western US forest wildfire activity. *Science* 313:940-943.

WHO (World Health Organization). 1986. *WHO Environmental Health Criteria 59: Principles for Evaluating Health Risks from Chemicals During Infancy and Early Childhood: The Need for a Special Approach*. Geneva: WHO Press.

WHO. 2004. Heat-waves: Risks and responses. Copenhagen, Denmark.

Whyatt RM, Garfinkel R, Hoepner LA, Andrews H, Holmes D, Williams MK, Reyes A, Diaz D, Perera FP, Camann DE, Barr DB. 2004. A biomarker validation study of prenatal chlorpyrifos exposure within an inner-city cohort during pregnancy. *Environmental Health Perspectives* 117(4):559-567.

Wu F, Biksey T, Karol M. 2007. Can mold contamination of homes be regulated? Lessons learned from radon and lead policies. *Environmental Science & Technology* 41(14):4861-4867.

Zhang K, Batterman S. 2009. Time allocated shifts and pollutant exposure due to traffic congestion: An analysis using the national human activity pattern survey. *The Science of the Total Environment* 407(21):5493-5500.

Zota A, Adamkiewicz G, Levy JI, Spengler JD. 2005. Ventilation in public housing: Implications for indoor nitrogen dioxide concentrations. *Indoor Air* 15(6):393-401.

3

Government and Private-Sector Involvement in Climate Change, Indoor Environment, and Health Issues

Several government and private-sector bodies are involved in various issues of climate change, indoor environment, and health. This chapter identifies them and summarizes their work in those issues. It also lists some major sources of data on the characteristics of buildings, the indoor environment, and public health and discusses how they might inform questions about the intersection between them.

FEDERAL GOVERNMENT AGENCIES AND DEPARTMENTS

The 2010 National Research Council report *Informing an Effective Response to Climate Change* lists 19 US federal executive and legislative branch bodies that are involved in or affected by decisions about climate change (NRC, 2010). This section lists the entities that are most directly involved in issues related to the intersection between climate change, the indoor environment, and health and identifies some of their work. Chapter 8 provides additional detail on programs related to building weatherization and energy efficiency.

US Environmental Protection Agency

The US Environmental Protection Agency (EPA)—the sponsor of the present study—conducts and coordinates research on a broad array of issues associated with climate change. Its purview includes both the outdoors and some indoor environments.[1] The bulk of EPA's efforts are directed toward research on and regulation of greenhouse gases, but the agency's Indoor Environments Division addresses climate-change questions as part of its objective to protect the public's health by promoting healthier indoor environments.

One major initiative is the ENERGY STAR voluntary building-certification program, which promotes the use of low-energy–demand designs, construction, and appliances. EPA cites lower greenhouse-gas emissions as one the benefits of certified homes (EPA, 2010e). The voluntary Indoor airPLUS standard allows builders who have already met ENERGY STAR requirements to apply an additional label to structures that have met criteria that include resistance to outdoor water intrusion, mitigation of opportunities for indoor dampness, a heating, ventilating, and air-conditioning (HVAC) system that meets American Society of Heating, Refrigerating, and Air-Conditioning Engineers standards for ventilation, and low-emission building materials (EPA, 2009b).

In late 2010, the agency released a draft of voluntary *Healthy Indoor Environment Protocols for Home Energy Upgrades* for public comment (EPA, 2010b). The protocols were developed in conjunction with the Department of Energy (DOE) *Workforce Guidelines for Home Energy Upgrades* (DOE, 2011) and focus on potential health effects of weatherization and other retrofits intended to promote energy efficiency. They touch on such issues as moisture, emissions from building materials, and ventilation and offer guidance on exposure assessment, mitigation, and adaptation strategies.

EPA specifically addresses the subject of the present report in an *Indoor Air Quality and Climate Readiness* Web site that in late 2010 included weatherization and indoor air-quality briefing material and links to more general indoor environmental-health information (EPA, 2010d). Several other information and education programs indirectly address building problems and exposures that have been associated with climate change and the indoor environment and with remediation of their adverse effects. The Agency's Tools for Schools program, for example, seeks to "prevent and solve the majority of indoor air problems with minimal cost and involvement" (EPA, 2009a, p. i).[2] As was the case with Indoor airPLUS, actions address outdoor-water intrusion, indoor dampness, proper ventilation, well-maintained HVAC systems, and low-emission building materials.

[1] Workplace environmental problems are under the jurisdiction of the Occupational Safety and Health Administration. This report touches on issues in offices but does not address industrial environments, which may also be adversely affected by climate change (Nilsson and Kjellstrom, 2010).

[2] These topics are also dealt with in the 2006 National Academies report *Green Schools: Attributes for Health and Learning* (NRC, 2006).

A cooperative agreement program announced in late 2010 disseminated $2.4 million to local government, educational institutions, and nonprofit organizations for "demonstration, training, education, and/or outreach projects that seek to reduce exposure to indoor air pollutants" and that would yield measurable results (EPA, 2010c).

EPA's Environmental Technology Verification (ETV) Program was initiated in 1995 to evaluate environmental technologies and make them readily available for the mass market for the benefit of the general public (EPA, 2011a). One of its main goals is to standardize testing among different companies and products. One such standardization was of the accuracy of technology that tests building pressure to determine whether contaminants in buildings are due to vapor intrusion or to other product emissions (ETV, 2010). Another initiative investigates microorganism-resistant building material for mold resistance, emissions of volatile organic compounds (VOCs) and aldehydes, and moisture content (RTI International, 2008).

EPA also partners with other federal agencies to conduct research. In collaboration with the Department of Housing and Urban Development (HUD), EPA conducted a national survey that measured allergens, including mold, and pesticides in homes (Stout et al., 2009). The data have since been used to examine the indoor environment and potential health risks to occupants. It also cochairs the Federal Interagency Committee on Indoor Air Quality with four other federal agencies.[3] This committee coordinates research and facilitates communication on indoor-air topics, including excessive dampness, mold, ventilation, emissions from building materials, and "green buildings."

National Institutes of Health

The National Institutes of Health (NIH) is the principal biomedical research arm of the Department of Health and Human Services (HHS). It conducts and sponsors investigations on a broad array of health topics and fosters both basic and applied research. Climate-change–related work at NIH falls principally under the aegis of the National Institute of Environmental Health Sciences (NIEHS), which holds primary responsibility for conducting and funding environmental health research. In 2010, that institute released the results of an effort by the Interagency Working Group

[3] The committee's Web site notes that "the CIAQ is co-chaired by EPA, the Consumer Product Safety Commission, the Department of Energy, the National Institute for Occupational Safety and Health, and the Occupational Safety and Health Administration. Other federal departments and agencies participate as members" (EPA, 2010a).

on Climate Change and Health[4] (NIEHS, 2010). The stated purpose of *A Human Health Perspective on Climate Change* was to (p. iv)

> identify research needs for all aspects of the research-to-decision making pathway that will help us understand and mitigate the health effects of climate change, as well as ensure that we choose the healthiest and most efficient approaches to climate change adaptation.

Among the research needs identified were studies addressing the health effects of indoor dust on asthma exacerbation, including changes in dust composition resulting from climate change (p. 15); how changes in temperature and precipitation affect exposure to toxic chemicals (p. 19); the effects of climate change on outbreak incidence, geographic range, and growth cycles of insect pests and pathogens that cause human disease (p. 27); risk factors for illness and death associated with acute exposure to extreme heat events and chronic exposure to increased average temperatures; and the health benefits of the use of environmental design principles to reduce the high thermal mass of urban areas (p. 31). The report also called for research aimed at anticipating, detecting, and responding to climate-change–induced and –exacerbated health problems and identifying vulnerable populations. In July 2010, NIH announced that it would operationalize those recommendations by providing research funding through a program intended to "examine the differential risk factors of populations that lead to or are associated with increased vulnerability to exposures, diseases and other adverse health outcomes related to climate change" (NIH, 2010).

NIEHS collaborates with EPA to support several Children's Environmental Health Research Centers, which conduct and support studies of the effects of environmental exposures. As noted later in this chapter, it cooperated with the Department of Housing and Urban Development's Office of Lead Hazard Control to conduct the National Survey of Lead and Allergens in Housing. The study gathered data on indoor allergen exposure that allowed NIEHS to "assess the magnitude of levels of indoor allergens in the United States housing stock" and "evaluate differences in population exposure to allergens based on factors such as region/geography, ethnicity, socioeconomic status, and housing type" (NIEHS, 2011).

[4] The working group comprised representatives of the Centers for Disease Control and Prevention, HHS's Office of the Secretary, EPA, the National Aeronautics and Space Administration, NIEHS, NIH's Fogarty International Center, the National Oceanic and Atmospheric Administration, the Department of State, the US Department of Agriculture, and the US Global Change Research Program.

Centers for Disease Control and Prevention

The Centers for Disease Control and Prevention (CDC), which also falls under the aegis of HHS, takes a public-health approach to climate-change–related work that includes (CDC, 2009b)

- Tracking data on environmental conditions, disease risks, and disease occurrence related to climate change.
- Expanding capacity for modeling and forecasting health effects that may be climate-related.
- Enhancing the science base to understand the relationship between climate change and health outcomes better.
- Identifying locations and population groups at greatest risk for specific health threats, such as heat waves.
- Communicating the health-related aspects of climate change, including risks and ways to reduce them, to the public, decision-makers, and health-care providers.

One component of the work is the Climate-Ready States and Cities Initiative. The initiative is intended to support health-department efforts to assess, plan for, and build capacity to respond to climate-change–related health effects (CDC, 2010a). Eight states[5] and two cities[6] were awarded grants totaling $5.25 million in 2010 to pursue projects. Many of them listed issues related to the indoor environment, such as heat-stress morbidity and mortality, as subjects to focus on, but indoor environmental quality does not appear to be among the concerns being addressed.

CDC's National Center for Environmental Health (NCEH) seeks to improve the nation's health status by avoiding diseases and disability caused by noncommunicable environmental factors (CDC, 2011a). It assigns high priority to vulnerable populations—specifically, children, the elderly, and people who have disabilities. NCEH's activities include lead-poisoning prevention and environmental-health workforce development and capacity-building. Its climate-change–related work includes prevention of carbon monoxide (CO) poisoning from home electricity generators during power outages.

CDC also collects surveillance data on diseases related to environmental changes via its National Environmental Public Health Tracking Network (CDC, 2010b). The network includes monitoring of home contaminants—as of 2010, lead and CO. Although it was not designed to investigate climate-change effects, the director of CDC's Division of Environmental

---

[5] Arizona, Maine, Massachusetts, Michigan, Minnesota, New York, North Carolina, and Oregon.

[6] New York City and San Francisco.

Hazards and Health Effects asserted in 2007 that it would be "an excellent tool" for such purposes (Late, 2007).

CDC has also provided funding for research projects on such topics as adverse exposures and health problems related to extreme weather events (Brandt et al., 2006; CDC, 2006).

National Institute for Occupational Safety and Health

CDC's National Institute for Occupational Safety and Health (NIOSH) examines the health consequences of occupational environments. NIOSH has conducted extensive research to evaluate the effects of indoor environments on occupant health and to characterize the factors that contribute to poor health outcomes. No specific research focuses on climate change and occupational-health issues, but the institute has investigated adverse respiratory health effects resulting from damp or water-damaged occupational environments (Cox-Ganser et al., 2005, 2009; Park et al., 2006) and has developed tools to assess indoor moisture to guide preventive actions. Investigators also examine the products of indoor chemistry[7] and their health effects (Anderson et al., 2010) and seek to determine the mechanisms by which indoor molds stimulate allergic responses (Green et al., 2009). NIOSH conducts research into the effects of "green" jobs on health, recognizing the new exposures and conditions associated with jobs designed to support activities that lead to energy efficiency and less environmental effect.

More generally, the potential for increased heat stress in indoor occupational environments has been flagged as a health and productivity issue in other countries (Kjellstrom et al., 2009a,b).

Department of Energy

DOE research activities include energy efficiency, clean-energy technology, and greenhouse-gas emission reduction. The department's Building Technologies Program does not identify climate change as a motivating factor but conducts work that addresses the topic through programs that seek to reduce energy demands and promote good indoor air quality. Research and development initiatives include support of revisions of ventilation and building codes; improvement of exposure-assessment, ventilation, filtering, and air-cleaning technologies; source reduction of VOCs; and better

[7] Indoor chemistry refers to the oxidation-reduction, acid-base, hydrolysis, decomposition, and other reactions that occur in indoors as a result of the interaction between various chemicals in the air, furnishing, floor and wall coverings, cleaning supplies and other constituents of the indoor environment.

understanding of the effects of energy-efficiency measures on health and productivity (DOE, 2010). Much of the work is conducted by the Indoor Environment Division of DOE's Lawrence Berkeley National Laboratory (LBNL, 2010). DOE's extensive work in weatherization and energy efficiency in buildings is addressed in Chapter 8.

Department of Housing and Urban Development

HUD's climate-change–related work focuses on the built environment and sustainable building practices—specifically measures to reduce energy consumption (HUD, 2010d). Among its efforts are Sustainable Communities Regional Planning Grants, which include predisaster mitigation plans and climate-change–impact assessments among the eligible activities (HUD, 2010b). HUD's Sustainable Communities Initiative promotes green building design and construction, but, although it mentions improved public health as a benefit of the program, that is not its focus (HUD, 2010c).

HUD also cooperates with DOE on the implementation of its Weatherization Assistance Program (WAP), identifying low-income properties (public housing, assisted housing, and others given special status under the enabling legislation) that are eligible for weatherization funds. A 2009 Memorandum of Understanding between the agencies streamlined the process for evaluating candidate properties for the program. HUD estimates that approximately 3 million housing units are potentially eligible for assistance (HUD, 2010a).

Federal Emergency Management Agency

FEMA, a part of the US Department of Homeland Security (DHS), has responsibility within the federal government to "build, sustain, and improve [the nation's] capability to prepare for, protect against, respond to, recover from, and mitigate all hazards" (FEMA, 2010c). This includes providing guidance on identifying and remediating problematic dampness and mold (FEMA, 2003), and responding to flood (FEMA, 2010a) and hurricane (FEMA, 2010b) damage. FEMA also collects and disseminates disaster epidemiology data and cooperates with agencies at all levels of government including Department of Homeland Security and Office for Interoperability and Compatibility in developing technical standards and specifications, and prioritizing emergency development.

The agency's Fiscal Years 2011–2014 strategic plan states that "challenges posed by climate change, such as more intense storms, frequent heavy precipitation, heat waves, drought, extreme flooding, and higher sea levels, have the potential to change significantly the types and magnitudes of hazards faced by communities and the emergency management professionals serving them" (FEMA, 2011). It has taken and is continuing to take several steps to respond to these challenges. These include a research effort initiated in 2009 to evaluate the potential effect of climate change on flood risk, and hence flood insurance (Lehmann, 2009).

US Global Change Research Program

The US Global Change Research Program (USGCRP) serves as the coordinating body for federal research on climate change and its effects on society (USGCRP, 2011). It comprises EPA, the Agency for International Development, the US Departments of Agriculture, the Department of Commerce,[8] the Department of Defense, DOE, HHS, the Department of State, the Department of the Interior, the Department of Transportation, the National Aeronautics and Space Administration, the National Science Foundation, and the Smithsonian Institution. The USGCRP has produced a series of reviews of scientific evidence, including a 2009 assessment of the state of scientific knowledge regarding global climate-change effects in the United States (USGCRP, 2009). The program maintains an Interagency Crosscutting Group on Climate Change and Human Health (CCHHG), but the present committee could not identify any work that it has published that explicitly addresses indoor environmental quality or building-related issues.

GOVERNMENT HOUSING AND HEALTH DATA COLLECTION

Various agencies and organizations conduct or sponsor studies that collect pieces of information useful in assessing the relationships between buildings, the environment, and health. Each of the existing surveillance systems noted below is designed to achieve specific goals related to buildings or public health, through, for example, monitoring of trends in pesticide use in homes, assessing the household costs of energy use, or examining changes in how people live and work in their buildings. The text below briefly summarizes the information that they collect and identifies potential opportunities and limitations in using them to assess potential effects of climate change on the indoor environment and occupant health. A thorough examination of methods and variables—a task beyond the scope of the present committee—would be needed to draw detailed conclusions concerning how to implement such a survey.

[8] Including the National Oceanic and Atmospheric Administration.

Housing and Building Surveys

American Housing Survey

The American Housing Survey (AHS) is conducted by the Census Bureau for HUD and includes apartments, single-family homes, mobile homes, housing characteristics, equipment, corresponding costs, and community characteristics, such as income and recent migration. The AHS is conducted in odd-numbered years and surveys the same housing units each time for comparison purposes. Every 6 years, specific data are collected on almost 50 metropolitan areas throughout the United States (US Census Bureau, 2008).

A substantial problem with the AHS from the standpoint of gathering information on the effects of climate change on indoor environments is that it is administered to the same housing unit every other year, whether or not the same residents live in the unit. Because the US population is relatively mobile, comparisons within this survey can be inconsistent (Acevedo-Garcia et al., 2004). Furthermore, renovations of a housing unit could have changed in ways that are material to the consideration of indoor environmental quality—for example, the purchase of a window air-conditioning unit or installation of new double-pane windows. A change of occupants of a housing unit would also mean changes in how the unit is used, which could influence and possibly confound variables used to evaluate indoor environmental quality.

American Healthy Homes Survey

EPA and HUD collected questionnaire and environmental data on a stratified, nationally representative sample of 1,131 US residences in 2005–2006 (Stout et al., 2009). Exposure measurements in the homes included pesticides, allergens, fungi, lead, and arsenic (Stout et al., 2009). The study built on a previous effort by HUD and NIEHS that measured lead and allergens in homes (Arbes et al., 2003; Cohn et al., 2004, 2006; Thorne et al., 2005). A future data collection planned to take place before 2020 will assess progress toward the *Healthy People 2020* goals regarding environmental exposures in noninstitutional US homes (Department of Health and Human Services, 2009).

Residential Energy Consumption Survey

The US Energy Information Administration conducts the Residential Energy Consumption Survey (RECS), a probability-sample survey that collects energy-related data on occupied primary housing units (Energy Information Administration, 2009). The first RECS was conducted in 1978; the most recent, in 2005, collected data on 4,381 households in housing units statistically selected to represent the 111 million housing units in the United States. Another wave of collection started in January 2011; its results are to be posted in late 2011 and early 2012 (Energy Information Administration, 2009). The collected data include physical characteristics, heating and cooling equipment, demographic characteristics of residents, and types of fuels used. Data are collected via three methods: in-person interviews with residents, in-person or telephone interviews with rental agents for units some or all of whose energy costs were included in the rent, and mail-in questionnaires from utility companies and suppliers.

Large Analysis and Review of European Housing and Health Status

In 2002–2003, the World Health Organization conducted the Large Analysis and Review of European Housing and Health Status (LARES), a cross-sectional survey to improve knowledge of the effects of housing on residents' physical well-being and mental health (Bonnefoy et al., 2007). Eight cities representing northern, southern, eastern, and western Europe participated. The sample in each city was randomly generated from resident registries, the local tax registry, or the national health insurance registry. LARES used three survey instruments: an inhabitant questionnaire that described residents' perceptions of their dwellings, a health questionnaire for inhabitants to report their health status (and that of children less than 12 years old), and a visual inspection by a trained surveyor (Bonnefoy et al., 2007). No physical measurements—such as temperature, humidity, and chemical or biologic exposures—were recorded. Teams of two technicians visited 3,373 dwellings and collected data on the health status of 8,519 inhabitants (Bonnefoy et al., 2007). LARES focused on such subjects as indoor air quality, noise effects, indoor dampness, and domestic accidents (WHO, 2011). The study examines indoor air environments and their connection to the building, but the data were centered on occupant perceptions of indoor air quality rather than on measurements, and climate-change–related factors were not assessed (WHO, 2011).

Health and Environment Surveys

National Health and Nutrition Examination Survey

The National Health and Nutrition Examination Survey (NHANES) is the most detailed large-scale survey of health status in the United States, with questionnaire, mental-health assessment, physical examination, laboratory, and some environmental data collected at home. The survey is national in scope and samples a representative population; it includes targeted "oversampling" to obtain sufficient data on various minority populations at different times.

The primary purpose of NHANES is to generate data that can be analyzed at a national level. However, coding schemes are available for researchers that provide information about subjects' locations by latitude and longitude, census tract and block, county, and state.

The survey could be enhanced in a number of ways to assess the effects of climate change. As Chapter 4 notes, there are outdoor air pollutants such as particulate matter and ozone whose levels may be affected by climate change, and outdoor levels influence indoor levels. One approach would be to collect valid, nationally representative air-toxics exposure data that could be linked in time and space to human health outcomes data. Previous important work in this field has been limited to community-level studies or the use of historical NHANES human health data linked to geographically interpolated air-toxics exposure data. The latter method has scientific value and has been used to support analyses of both NHANES and US National Health Interview Survey (NHIS) data. However, it has limitations, and there is a need for improved data collection.

NHANES also has the ability to measure concentrations of a wide variety of specific chemicals in blood and urine, and it does this for a number of environmental analytes of interest—such as lead, mercury, and organochlorines. In the past, blood concentrations of VOCs were also measured. It is therefore a primary source of national-level environmental-health data on the United States. NHANES has already conducted environmental sampling in homes during one cycle (2005–2006), and this could be repeated and expanded. Data collected included dust concentrations of dust mite, cockroach, dog, cat, rat, mouse, *Alternaria*, and *Aspergillus* allergens and serum concentrations of IgE antibodies to these antigens (Gergen et al., 2009; Visness et al., 2009). A summary of the NHANES environmental-health data is published in CDC's *National Report on Human Exposure to Environmental Chemicals* (CDC, 2011b) and in numerous peer-reviewed journal articles. NHANES also has the ability to perform direct air toxin exposure monitoring of individual participants for short periods (24–48 hours), but this data collection requires more extended efforts and costs than local environmental monitoring. Data on VOCs in the breathing zone of participants were collected from 2005 to 2010 (CDC, 2009a).

National Health Interview Survey

NHIS is a multistage probability-sample survey conducted by CDC's National Center for Health Statistics (CDC, 2009c). It reaches 75,000–100,000 persons in the United States each year and collects a wide array of sociodemographic and health information through direct visits to households. NHIS is considered the principal source of data on US asthma prevalence. Modular units collect data on subjects of special interest and could be designed to evaluate climate-change effects on health.

Behavioral Risk Factor Surveillance System

CDC's Behavioral Risk Factor Surveillance System (BRFSS) is a state-based data-collection effort that uses telephone surveys to obtain information (CDC, 2010c). It has been in operation since 1985 and conducts more than 400,000 interviews each year. As the name suggest, BRFSS focuses on how people conduct their daily lives and how this influences their health. It has several potential advantages as an instrument for amassing climate-change and health information. The survey is already being used to examine data at geographic region, state, and local levels, and its Selected Metropolitan/Micropolitan Area Risk Trends (SMART) database allows breakouts of information on more than 200 metropolitan and micropolitan statistical areas. BRFSS also allows states to add questions to suit local needs or to assess the effects of particular events, such as hurricanes.

National Environmental Public Health Tracking Program

The National Environmental Public Health Tracking Program, developed by CDC, is coordinating a national system to track environmental hazards and related diseases (CDC, 2007). The program updates traditional surveillance systems with geographic information systems (GIS). Many of the data arise from state and local health-department grantees. In addition, CDC has collaborated with several other federal agencies and professional organizations to provide data for the program. The National Environmental Public Health Tracking Network promotes information-system standards to integrate local, state, and national databases on environmental hazards, environmental exposures, and health effects, including outdoor and indoor air exposures.

Other Data Sources

Calculating past exposures and modeling trends in outdoor air pollution would require the use of other existing databases. EPA and various state agencies have historical databases of air-quality measurements and exposure assessments, including the EPA National Air Toxics Assessments (EPA, 2011b) and the California Air Quality Resources Board air quality monitoring databases (Cal/EPA ARB, 2011a).

Existing health surveys are not, for the most part, designed to assess major time-limited events, such as hurricanes and heat waves. Syndromic surveillance[9] systems may prove useful for assessing health effects of these events, but not all are structured to do so.[10] In 2010, CDC initiated a redesign of *BioSense*—which was created in 2003 to "establish an integrated national public health surveillance system for early detection and rapid assessment of potential bioterrorism-related illness" (CDC, 2011c)—to imbed it more firmly in state and local public-health systems and make it more responsive to local needs for information.

Synthesis—Surveillance Systems to Track Climate-Change Effects on Indoor Environmental Quality and Health

To track the effects of climate change on indoor environmental quality and health, it will be important to gather information over time and in specific geographic regions to assess variations and the different effects associated with them. Environmental and building factors of interest include

- Outdoor temperature, humidity, and rainfall.
- Outdoor air quality, including levels of hazardous air pollutants and particulates.
- Building type or use—single-family residence, multihousing unit, school, office, or commercial space.
- Building and indoor environmental characteristics, including presence, type, and condition of HVAC system; air-exchange rate; building age, location, and setting (urban, suburban, or rural); temperature, humidity, allergens, and chemical contamination; and mitigation strategies implemented.

Health outcomes of interest include

- Asthma—prevalence and severity.
- Allergies—prevalence and severity.
- Vectorborne illness.
- Waterborne illnesses.
- Reproductive outcomes.
- Cancer.

[9] Syndromic surveillance "is concerned with continuous monitoring of public health-related information sources and early detection of adverse disease events" (Yan et al., 2008). These include epidemics and bioterrorism incidents.

[10] Chen et al. (2010) describe syndromic surveillance systems that have been used for one-time special and large-scale events; Josseran et al. (2010) relate the use of a system in France to monitor the effects of a heat wave.

- Heat stress.
- Excess mortality during periods of excessive heat or extreme weather events.

A 2009 report from the State Environmental Health Indicators Collaborative (SEHIC) listed several environmental-health indicators that are currently tracked, need validation, or need to be tracked (English et al., 2009). They included wildfires, pollen, temperature and humidity, drought, harmful algal blooms, and respiratory and allergic disease related to air quality. In addition, a report from the Interagency Climate Change Adaptation Task Force (ICCATF)—under the auspices of the White House Council on Environmental Quality—set a policy goal of coordinating capabilities of the federal government to support climate-change adaptation (2010). The ICCATF recommended that "agencies should work individually, collaboratively, and with the Task Force to ensure that resources are allocated to maximize their impact and avoid unnecessary duplication" (2010). Therefore, merely adding health-related indicators to housing surveys or adding housing-related indicators to health surveys could improve the tracking of climate-change effects but might result in redundancy. Because of the aforementioned limitations of surveillance systems and the lack of a consistent timeframe for measurement of the indicators (for example, some surveys are repeated every year and some sporadically, and some surveys ask about exposure or health outcomes in the preceding 3 months and some about them in the preceding year), combining datasets can be complicated at best and misleading at worst.

The ideal surveillance for assessing climate-change effects of indoor environment exposures and related health effects would be a national study with this clear focus. A model would be the National Children's Study (NCS), which will be prospective (that is, allow clear identification of trends in a given population), large (100,000 children), and representative of the population and will incorporate objective measurements of environmental exposures (including biomarkers) and health outcomes (Landrigan et al., 2006). An expansion of the NCS to include health outcomes of other members living in the household could be rather expensive, but it would leverage the existing environmental measurements from the home and enable followup of middle-age and older adults who could also be at risk for the effects of climate change.

Another option, albeit more limited, is the use of information technology in buildings and in assessment of health outcomes. Data loggers can easily track temperature and humidity, and recent advances have led to real-time measurement devices for environmental exposure (such as chemical and particles) and biomarker monitoring that can be coupled with accelerometers to track exposures when study participants wear them

(NIEHS, 2008). Substantial cost savings can be realized by bypassing the requirement of trained technicians to set up bulky equipment in the home, work, or school environment. In addition, the source apportionment of exposures could be refined while the ability to link biologically relevant exposures to health outcomes was improved The passive collection of data in buildings and from surveillance of participants at multiple times is possible only with improved information technology. However, the data gained will inevitably lead to better surveillance of climate-change–related exposures and health effects.

STATE AND LOCAL GOVERNMENTS

State and local government climate-change initiatives generally focus on greenhouse-gas emissions, energy efficiency, and preservation of infrastructure—including the public-health infrastructure—in the event of extreme weather or flooding. Among the ones that include consideration of the indoor environment and public health, the issues that are most often addressed are extreme heat and problems that disproportionately affect the elderly, low-income, and other vulnerable populations. Exposures in the indoor environment and health-related adaptation and mitigation efforts for buildings either are not addressed or are mentioned only in passing. A few examples are provided below.

Separately, there are isolated efforts on the state and local levels to address indoor environmental health concerns. Although they were not motivated by climate-change concerns, they address exposures that might be exacerbated by changing outdoor conditions. One, a California standard on emissions from building materials, is summarized below. A 2010 white paper by Levin provides details on that standard and voluntary standards in the United States that address emissions from building materials and products that may affect indoor environmental quality. The 2004 Institute of Medicine report *Damp Indoor Spaces and Health* compares the guidance on mold remediation offered by federal, local government, and private sources available at the time of its publication.

State Initiatives

The Pew Center on Global Climate Change reported that 36 states had comprehensive climate action plans in February 2011 and two more had plans under development (Pew, 2011a). They vary widely in scope and focus, but their building-sector initiatives typically include green-building standards, residential and commercial energy-conservation codes, and appliance-efficiency standards (Pew, 2011b).

California is among the states that offer specific regulatory guidance regarding environmental and public-health considerations for buildings. Its Specification 01350 establishes goals and provides guidelines for energy and material use in buildings; indoor air quality, including nontoxic performance standards for cleaning and maintenance products; and other occupant health and sustainability considerations (CalRecycle, 2011).

Specification 01350 includes provisions for evaluating VOC emissions from indoor sources. The testing is intended to limit health effects of exposure to VOCs and occurs at multiple stages during construction. It evaluates emission data on large-surface-area materials by using standard exposure scenarios for estimating VOC emissions and area-specific air flow rates. Specific VOCs are considered as separate pollutants to estimate possible health effects on building occupants more accurately. That means of measuring VOCs and indoor air quality has since been incorporated into sections of the draft International Green Construction Code and is influencing other green-building certification and labeling schemes (Levin, 2010). Levin's 2010 EPA white paper addresses Specification 01350 in greater detail.

The state's Green Building Standards Code (CalGreen) is intended to improve public health and safety through planning and design, energy efficiency, water efficiency and conservation, material conservation and resource efficiency, and environmental quality measures (California Building Standards Commission, 2010). The codes apply to state-regulated and owned buildings and structures, including public elementary and secondary schools and California State University buildings, as well as other buildings such as low-rise residential buildings and acute care hospitals and clinics (California Building Standards Commission, 2010). Among its provisions are mandatory measures that require low-emitting materials and coatings (based on Specification 01350 and other limits) and voluntary measures regarding indoor air quality.

The California Environmental Protection Agency's Air Resources Board has regulatory authority to evaluate and control air toxics under the state's 1983 Toxic Air Contaminant Identification and Control Act (Cal/EPA ARB, 2009). In 2009, that authority was used to promulgate an airborne toxic control measure to reduce formaldehyde emissions from composite wood products used in home construction, finishing, and furniture (Cal/EPA ARB, 2011).

City Initiatives

Many larger cities have or are developing climate-action plans, typically centered on infrastructure protection in coastal areas. Almost all public-health departments have plans in place to deal with heat-wave emergencies. Two of the more comprehensive efforts are summarized briefly below.

In 2008, New York City used a grant from the Rockefeller Foundation to establish a Panel on Climate Change as part of a larger effort to establish a long-term sustainability plan. The panel released a report in 2010 that took a risk-management approach to adaptation questions. It included a series of climate-change–related considerations that the authors believed should be taken in account in revising infrastructure design and performance standards, such as those for buildings (NYC Panel on Climate Change, 2010). Four primary hazards were identified—coastal flooding and storm surge, inland flooding, heat waves, and extreme wind events—all of which are also addressed in the city's natural-hazard mitigation plan (NYC Office of Emergency Management, 2009).

Chicago's Climate Change Action Plan includes the promotion of building design, construction, and operation practices that enhance energy efficiency and human health outcomes. The city requires that new government buildings conform to LEED Silver certification standards (Chicago Climate Task Force, 2008). The urban heat-island effect is a concern for the city, which experienced an extreme heat event in 1995 that resulted in more than 400 deaths in excess of the number otherwise expected (CDC, 1995; Kaiser et al., 2007). The action plan mentions that steps will be taken to identify at-risk populations and promote innovation to ameliorate heat islands, but it offers no specifics. A private-sector initiative, the Chicago Community Loan Fund, provides low- and middle-income housing financing and encourages the use of energy-efficient building standards and nontoxic and low-emission materials in the design and construction of affordable housing (Chicago Community Loan Fund, undated).

INTERGOVERNMENTAL PANEL ON CLIMATE CHANGE

The Intergovernmental Panel on Climate Change was created under the auspices of the UN Environment Programme and the World Meteorological Organization to review and assess research and information on climate change to enhance worldwide understanding of the topic (IPCC, 2010). Discussion of indoor air quality issues in its fourth report, which was published in 2007, focused on indoor biomass combustion and its adverse effects on human health (Metz et al., 2007). The report called indoor air pollution "a key environmental and public health peril for countless of the world's poorest, most vulnerable people" and advocated the adoption of cleaner-burning cooking stoves both to prevent health problems and to limit greenhouse-gas emissions. For developed countries, it noted that "the diffusion of new technologies for energy use and/or savings in residential and commercial buildings contributes to an improved quality of life and increases the value of buildings" (Metz et al., 2007). A fifth report was under development in early 2011. It will emphasize socioeconomic vulnerability to the effects of climate change and implications of sustainable development and risk management (IPCC, 2010).

PRIVATE SECTOR

The private sector plays a considerable role in issues of climate change, the indoor environment, and health. A few examples are listed below. White papers commissioned by EPA in support of the present study provided detailed information on industry and professional-organization initiatives regarding building materials and product-testing regimens (Levin, 2010), green-building rating systems (Srebric, 2010), and energy-conservation codes for commercial and residential buildings (Mudarri, 2010). All those are discussed elsewhere in this report.

American Society of Heating, Refrigerating and Air-Conditioning Engineers

The American Society of Heating, Refrigerating and Air-Conditioning Engineers (ASHRAE) is a professional organization that serves to advance the science of sustainable heating, ventilating, refrigeration, and air conditioning (ASHRAE, 2011). Its membership is drawn from private-sector, academic, and government professionals. ASHRAE has considerable involvement in indoor air-quality issues, in particular through standards[11] that it and the American National Standards Institute (ANSI) have developed for proper ventilation of commercial and residential buildings and the maintenance of thermal comfort in buildings. The standards, although voluntary and advisory, have been adopted into many building codes. The organization has also published the *Indoor Air Quality Guide*, which offers design and construction strategies to improve indoor air quality that go beyond those specified in codes and standards (ASHRAE, 2009b). ASHRAE's involvement in climate-change issues includes its *GreenGuide*—which provides information on sources of green design, construction, and operation practices (2010)—and a 2009 climate-change position document focused on reducing building emissions of greenhouse gases (2009a).

[11] *Standard 62: Ventilation for Acceptable Indoor Air Quality* and *Standard 55: Thermal Environmental Conditions for Human Occupancy*. A third document addressing how ventilation, the thermal environment, and other building characteristics jointly influence indoor environmental quality—*Guideline 10: Interactions Affecting the Achievement of Acceptable Indoor Environments*—was under development in early 2011.

LEED

LEED (Leadership in Energy and Environmental Design)—a component of the US Green Building Council, a building-trades association—promulgates voluntary certification standards for buildings that emphasize the reduction of climate-change effects. The standards include consideration of indoor air quality, but they focus primarily on increasing buildings' water and energy efficiency and decreasing their greenhouse-gas emissions and other aspects of their environmental footprint. Chapter 8 addresses LEED standards in greater detail.

Insurance Industry

A 2008 Ernst & Young study identified potential climate change as the greatest strategic risk facing the property and casualty insurance industry (Ernst & Young, 2008). Segments of the industry have been heavily involved in climate-change issues, particularly those related to reinsurance[12] (Nutter, 2010). The firm Swiss Re has published reports on the topic, addressing primarily the vulnerability of buildings and other infrastructure to catastrophic weather events (Swiss Re, 2002, 2010). Munich Re maintains *NatCatSERVICE*, which it characterizes as the most comprehensive global-loss database and which tracks the incidence of hurricanes, heat waves, flash floods, and other extreme weather events as part of a larger effort in cataloging natural catastrophes (Munich Re, 2003). There is a small literature on the effect of climate change on the insurance industry's business (Mills, 2005, 2007).

American Red Cross

American Red Cross emergency response and disaster preparedness programs offer relief and development assistance to millions of people annually who are affected by natural disasters. Their emergency response programs provide financial assistance to stimulate the local economy; relief supplies such as food, shelter materials, and hygiene kits; and trained volunteers who assess needs and implement critical relief services (ARC, 2011). The Red Cross works closely with FEMA to assist the US government agencies and community organizations in planning, coordinating, and providing mass care services for communities influenced by disasters (ARC, 2010).

---

[12] Reinsurance, simply put, is insurance that insurance companies take out to protect themselves against the risk of unusually large or numerous payouts on policies that they write. Reinsurance can become important when catastrophic events occur, especially if there is an anomalous number of them during a relatively short period.

Disaster epidemiology data developed by Red Cross/Red Crescent societies are used by government and other bodies for policy and planning purposes. Internationally, the Red Cross/Red Crescent Climate Centre concentrates on the humanitarian effects of climate change and extreme weather events. The Centre's mission is to educate and advocate for disaster risk reduction and climate adaptation; analyze relevant weather forecast data on all timescales; and incorporate understanding of climate risks into Red Cross/Red Crescent strategies, plans and procedures (RC/RCCC, 2011).

OBSERVATIONS

The preceding sections illustrate a fundamental problem. Multiple parts of government and the private sector have a stake in issues of climate change, indoor environmental quality, and public health, but no one body has assumed or attempted to assume the lead responsibility. As a result, there is a lack of leadership in identifying potential hazards, formulating solutions, and setting research and policy priorities.

The present report cannot solve that problem. Its aim is instead to highlight important issues for decision-makers and the scientific community. In approaching that aim, it seeks to draw special attention to

- Ways in which the information needed to make informed decisions is lacking.
- Ways in which initiatives aimed at reducing climate-change risks have the potential to inadvertently exacerbate problems in the indoor environment.
- How it may be possible to achieve a healthier indoor environment at lower cost, with lower emissions, or both than is currently the case.

REFERENCES

Acevedo-Garcia D, Osypuk TL, Werbel RE, Meara ER, Cutler DM, Berkman LF. 2004. Does housing mobility policy improve health? *Housing Policy Debate* 15(1):49-98.

Anderson SE, Jackson LG, Franko J, Wells JR. 2010. Evaluation of dicarbonyls generated in a simulated indoor air environment using an in vitro exposure system. *Toxicological Sciences* 115(2):453-461.

Arbes SJ, Cohn RD, Yin M, Muilenberg ML, Burge HA, Friedman W, Zeldin DC. 2003. House dust mite allergen in US beds: Results from the First National Survey of Lead and Allergens in Housing. *The Journal of Allergy and Clinical Immunology* 111(2):408-414.

ARC (American Red Cross). 2010. *FEMA, Red Cross to share mass care responsibility for U.S. emergencies.* http://www.redcross.org/portal/site/en/menuitem.1a019a978f421296e81ec89e43181aa0/?vgnextoid=988d4c642b2db210VgnVCM10000089f0870aRCRD (accessed April 17, 2011).

ARC. 2011. *Emergency disaster response and preparedness*. http://www.redcross.org/portal/site/en/menuitem.d229a5f06620c6052b1ecfbf43181aa0/?vgnextoid=cc0795e5ded8e110VgnVCM10000089f0870aRCRD&vgnextchannel=5002af3fbac3b110VgnVCM10000089f0870aRCRD (accessed April 17, 2011).

ASHRAE (American Society of Heating, Refrigerating and Air Conditioning Engineers). 2009a. *ASHRAE position document on climate change*. http://www.ashrae.org/File%20Library/docLib/Public/20060823_2006710123326_347.pdf (accessed February 21, 2011).

ASHRAE. 2009b. *Indoor air quality guide—Best practices for design, construction, and commissioning*. Atlanta, GA: ASHRAE.

ASHRAE. 2010. *ASHRAE GreenGuide: The design, construction, and operation of sustainable buildings*. Atlanta, GA: ASHRAE.

ASHRAE. 2011. *About us*. http://www.ashrae.org/aboutus/ (accessed February 21, 2011).

Bonnefoy X, Braubach M, Davidson M, Robbel N. 2007. A pan-European housing and health survey: Description and evaluation of methods and approaches. *International Journal of Environment and Pollution* 30(3/4):363-383.

Brandt M, Brown C, Burkhart J, Burton N, Cox-Ganser J, Damon S, Falk H, Fridkin S, Garbe P, McGeehin M, Morgan J, Page E, Rao C, Redd S, Sinks T, Trout D, Wallingford K, Warnock D, Weissman D. 2006. Mold prevention strategies and possible health effects in the aftermath of hurricanes and major floods. *MMWR* 55(RR08):1-27.

Cal/EPA (California Environmental Protection Agency). 2009. *ARB mission and goals*. http://www.arb.ca.gov/html/mission.htm (accessed May 1, 2011).

Cal/EPA ARB (Air Resources Board). 2009. *Consumer products and services—California's air toxics program*. http://www.arb.ca.gov/html/brochure/airtoxic.htm (accessed April 28, 2011).

Cal/EPA ARB. 2011. *Air quality monitoring network*. http://www.arb.ca.gov/aqd/aqmoninca.htm (accessed March 15, 2011).

California Building Standards Commission. 2010. *2010 California Green Building Standards Code: California Code of Regulations, Title 24, Part 11*. Sacramento, CA.

CalRecycle (California Department of Resources Recycling and Recovery). 2011. *Sustainable (Green) Building Section 01350*. http://www.calrecycle.ca.gov/greenbuilding/specs/section01350/ (accessed February 22, 2011).

CDC (Centers for Disease Control and Prevention). 1995. Heat-related mortality—Chicago, July 1995. *MMWR Morbidity and Mortality Weekly Report* 44:577-579.

CDC. 2006. Health concerns associated with mold in water-damaged homes after Hurricanes Katrina and Rita—New Orleans area, Louisiana, October 2005. *MMWR* 55(02):41-44.

CDC. 2007. *The National Environmental Public Health Tracking Program: Healthy informed communities 2007*. http://www.cdc.gov/nceh/tracking/pdfs/aag07.pdf (accessed January 31, 2011).

CDC. 2009a. *About the National Health Nutrition and Examination Survey*. http://www.cdc.gov/nchs/nhanes/about_nhanes.htm (accessed March 15, 2011).

CDC. 2009b. *CDC policy on climate change and public health*. http://www.cdc.gov/climatechange/pubs/Climate_Change_Policy.pdf (accessed October 27, 2010).

CDC. 2009c. *About the National Health Interview Survey*. http://www.cdc.gov/nchs/nhis/about_nhis.htm (accessed March 15, 2011).

CDC. 2010a. *CDC's climate-ready states & cities initiative*. http://www.cdc.gov/climatechange/climate_ready.htm (accessed October 27, 2010).

CDC. 2010b. *National environmental public health tracking*. http://cdc.gov/nceh/tracking/pib.htm. (accessed September 16, 2010).

CDC. 2010c. *Health risks in the United States. Behavioral risk factor surveillance system: At a glance 2010.* http://www.cdc.gov/chronicdisease/resources/publications/AAG/brfss.htm (accessed March 15, 2011).

CDC. 2011a. *National Center for Environmental Health.* http://www.cdc.gov/nceh/ (accessed February 27, 2011).

CDC. 2011b. *National report on human exposure to environmental chemicals.* http://www.cdc.gov/exposurereport/ (accessed March 15, 2011).

CDC. 2011c. *Biosense.* http://www.cdc.gov/biosense/ (accessed July 15, 2011).

Chen H, Zeng D, Yan P. 2010. *Infectious disease informatics: Syndromic surveillance for public health and bio-defense.* New York: Springer Science & Business Media, LLC.

Chicago Climate Task Force. 2008. *Chicago climate action plan—our city, our future.* http://www.chicagoclimateaction.org/filebin/pdf/finalreport/CCAPREPORTFINALv2.pdf (accessed November 30, 2010).

Chicago Community Loan Fund (undated). *Building for sustainability creating energy-efficient and environmentally friendly affordable housing in Chicago.* http://www.swaraj.org/shikshantar/chicagogreen.pdf (accessed November 30, 2010).

Cohn RD, Arbes SJ, Yin M, Jaramillo R, Zeldin DC. 2004. National prevalence and exposure risk for mouse allergen in US households. *The Journal of Allergy Clinical Immunology* 113(6):1167-1171.

Cohn RD, Arbes SJ, Jaramillo R, Reid LH, Zeldin DC. 2006. National prevalence and exposure risk for cockroach allergen in U.S. households. *Environmental Health Perspectives* 114(4):522-526.

Cox-Ganser JM, White S, Jones R, Hilsbos K, Storey E, Enright P, Rao CY, Kreiss K. 2005. Respiratory morbidity in office workers in a water-damaged building. *Environmental Health Perspectives* 113(4):485-490.

Cox-Ganser JM, Rao CY, Park JH, Schumpert JC, Kreiss K. 2009. Asthma and respiratory symptoms in hospital workers related to dampness and biological contaminants. *Indoor Air* 19(4):280-290.

Department of Health and Human Services. 2009. *Healthy People 2020: The road ahead.* http://www.healthypeople.gov/hp2020 (accessed January 10, 2011).

DOE (US Department of Energy). 2010. *Indoor air quality R&D.* http://www1.eere.energy.gov/buildings/indoor_air.html (accessed November 30, 2010).

DOE. 2011. *Residential retrofit guidelines.* http://www1.eere.energy.gov/wip/retrofit_guidelines.html (accessed February 27, 2011).

Energy Information Administration. 2009. *Residential Energy Consumption Survey (RECS).* http://www.eia.doe.gov/emeu/recs (accessed January 11, 2011).

English PB, Sinclair AH, Ross Z, Anderson H, Boothe V, Davis C, Ebi K, Kagey B, Malecki K, Schultz R, Simms E. 2009. Environmental health indicators of climate change for the United States: Findings from the State Environmental Health Indicator Collaborative. *Environmental Health Perspectives* 117(11):1673-1681.

EPA (Environmental Protection Agency). 2009a. *Indoor air quality tools for schools. Reference guide.* http://www.epa.gov/iaq/schools/pdfs/kit/reference_guide.pdf (accessed October 29, 2010).

EPA. 2009b. *Indoor airPLUS construction specifications.* http://www.epa.gov/indoorairplus/pdfs/construction_specifications.pdf (accessed October 29, 2010).

EPA. 2010a. *Federal interagency committee on indoor air quality.* http://www.epa.gov/iaq/ciaq/ (accessed November 17, 2010).

EPA. 2010b. *Healthy indoor environment protocols for home energy upgrades.* http://www.epa.gov/iaq (accessed November 18, 2010).

EPA. 2010c. *Indoor air quality.* http://www.epa.gov/iaq (accessed September 15, 2010).

EPA. 2010d. *Indoor air quality and climate readiness.* http://www.epa.gov/iaq/climate readiness/ (accessed November 17, 2010).

EPA. 2010e. *Join the fight against climate change: Take the Energy Star pledge: Energy Star.* http://www.energystar.gov/index.cfm?fuseaction=globalwarming.showPledgeHome (accessed October 29, 2010).

EPA. 2011a. *Environmental technology verification program: Basic information.* http://www.epa.gov/nrmrl/std/etv/basic.html (accessed February 27, 2011).

EPA. 2011b. *National air toxics assessments.* http://www.epa.gov/ttn/atw/natamain/ (accessed March 15, 2011).

Ernst & Young. 2008. *Strategic business risk—Insurance.* http://aaiard.com/11_2008/2008_Strategic_Business_Risk_-_Insurance.2.pdf (accessed February 20, 2011).

ETV (Environmental Technology Verification Program). 2010. *Quality assurance project plan for verification of building pressure control for the assessment of vapor intrusion.* Columbus, OH: Battelle.

FEMA (Federal Energy Management Agency). 2003. *Dealing with mold & mildew in your flood damaged home.* http://www.fema.gov/pdf/reg-x/mold_mildew.pdf.

FEMA. 2010a. *Flood response.* http://www.fema.gov/hazard/midwestfloods.shtm (accessed April 26, 2011).

FEMA. 2010b. *Hurricane.* http://www.fema.gov/hazard/hurricane/index.shtm (accessed April 26, 2011).

FEMA. 2010c. *Protecting your home and property from flood damage. Mitigation ideas for reducing flood loss.* http://www.fema.gov/library/file?type=originalAccessibleFormatFile&file=protecting_home_book_508compliant.pdf&fileid=c37fe0c0-615f-11e0-b6f6-001cc4568fb6 (accessed April 26, 2011).

FEMA. 2011. *FEMA Strategic Plan Fiscal Years 2011-2014*; FEMA P-806 / February 2011. http://www.fema.gov/txt/about/strategic_plan11.txt (accessed April 17, 2011).

Gergen PJ, Arbes SJ Jr, Calatroni A, Mitchell HE, Zeldin DC. 2009. Total IgE levels and asthma prevalence in the US population: Results from the National Health and Nutrition Examination Survey 2005-2006. *Journal of Allergy and Clinical Immunology* 124(3): 447-453.

Green BJ, Tovey ER, Beezhold DH, Perzanowksi MS, Acosta LM, Divjan AI, Chew GL. 2009. Surveillance of fungal allergic sensitization using the fluorescent halogen immunoassay. *Journal de Mycologie Medicale* 19(4):253-261.

HUD (US Department of Housing and Urban Development). 2010a. *Fact sheet—HUD-DOE weatherization partnership. Streamlining weatherization assistance in affordable housing.* http://portal.hud.gov/hudportal/documents/huddoc?id=factsheet_doe_weatherize_3.pdf (accessed April 26, 2011).

HUD. 2010b. *Notice of funding availability (NOFA) for HUD's fiscal year 2010 sustainable communities regional planning grant program.* http://www.hud.gov/offices/adm/grants/nofa10/scrpgsec.pdf (accessed October 28, 2010).

HUD. 2010c. *Sustainable communities regional planning grants.* http://portal.hud.gov/portal/page/portal/HUD/program_offices/sustainable_housing_communities/Sustainable%20Communities%20Regional%20Planning%20Grants (accessed September 30, 2010).

HUD. 2010d. *Sustainable housing and communities.* http://portal.hud.gov/portal/page/portal/HUD/program_offices/sustainable_housing_communities (accessed September 18, 2010).

Interagency Climate Change Adaptation Task Force. 2010. *Progress report of the Interagency Climate Change Adaptation Task Force: Recommended actions in support of a national climate change adaptation strategy.* Washington, DC: The White House Council on Environmental Quality.

IPCC (Intergovernmental Panel on Climate Change). 2010. *Organization.* http://www.ipcc.ch/organization/organization.shtml (accessed January 25, 2011).

Josseran L, Fouillet A, Caillère N, Brun-Ney D, Ilef D, Brucker G, Medeiros H, Astagneau P. 2010. Assessment of a syndromic surveillance system based on morbidity data: Results from the Oscour network during a heat wave. *PLoS One* 5(8):e11984.

Kaiser R, Le Tertre A, Schwartz J, Gotway CA, Daley WR, Rubin CH. 2007. The effect of the 1995 heat wave in Chicago on all-cause and cause-specific mortality. *American Journal of Public Health* 97(Suppl 1):S158-S162.

Kjellstrom T, Holmer I, Lemke B. 2009a. Workplace heat stress, health and productivity—an increasing challenge for low and middle-income countries during climate change. *Global Health Action* 2:2-6.

Kjellstrom T, Kovats RS, Lloyd SJ, Holt T, Tol RS. 2009b. The direct impact of climate change on regional labour productivity. *Archives of Environmental & Occupational Health* 64(4):217-227.

Landrigan PJ, Trasande L, Thorpe LE, Gwynn C, Lioy PJ, D'Alton ME, et al. 2006. The National Children's Study: A 21-year prospective study of 100,000 American children. *Pediatrics* 118(5):2172-2186.

Late M. 2007. U.S. environmental public health tracking programs gain success: Partners working on nationwide network. *Nation's Health* 37(5):21.

LBNL (Lawrence Berkeley National Laboratory). 2010. *Indoor environment division*. http://eetd.lbl.gov/IEP/IEP.html/ (accessed November 30, 2010).

Lehmann E. 2009. FEMA launches effort to measure impact of climate change on flood insurance. *New York Times*, June 11, 2009.

Levin H. 2010. *National programs to assess IEQ effects of building material and products*. Washington, DC: EPA Indoor Environments Division. http://www.epa.gov/iaq/pdfs/hal_levin_paper.pdf (accessed February 18, 2011).

Metz B, Davidson OR, Bosch PR, Dave R, Meyer LA. 2007. *Contribution of working group III to the fourth assessment report of the Intergovernmental Panel on Climate Change, 2007*. New York: Cambridge University Press.

Mills E. 2005. Insurance in a climate of change. *Science*. 309:1040-1044.

Mills E. 2007. Synergisms between climate change mitigation and adaptation: An insurance perspective. *Mitigation and Adaptation Strategies in Global Change*. 12:809-842.

Mudarri D. 2010. *Building codes and indoor air quality*. Washington, DC: EPA Indoor Environments Division.

Munich Re. 2003. *NatCatSERVICE® A guide to the Munich Re database for natural catastrophes*. Munich, Germany: Münchener Rückversicherungs-Gesellschaft.

NIEHS (National Institute of Environmental Health Sciences). 2008. *NIEHS Exposure Biology Program: A component of the NIH Genes, Environment and Health Initiative (GEI)*. http://www.niehs.nih.gov/health/docs/expbio-gei-fact.pdf (accessed January 11, 2011).

NIEHS. 2010. *A human health perspective on climate change*. http://www.niehs.nih.gov/health/docs/climatereport2010.pdf (accessed November 22, 2010).

NIEHS. 2011. *National Survey of Lead and Allergens in Housing (NSLAH)*. http://www.niehs.nih.gov/research/clinical/join/studies/riskassess/nslah.cfm (accessed May 1, 2011).

NIH (National Institutes of Health). 2010. *Climate change and health: Assessing and modeling population vulnerability to climate*. http://grants.nih.gov/grants/guide/pa-files/PAR-10-235.html (accessed November 22, 2010).

Nilsson M, Kjellstrom T. 2010. Climate change impacts on working people: How to develop prevention policies. *Global Health Action* 3:5774 - DOI: 10.3402/gha.v3i0.5774.

NRC (National Research Council). 2006. *Green schools: Attributes for health and learning*. Washington, DC: The National Academies Press.

NRC. 2010. *Informing an effective response to climate change*. Washington, DC: The National Academies Press.

Nutter F. 2010. *Climate change and the built environment.* Presentation before the Committee on the Effect of Climate Change on Indoor Air Quality and Public Health, June 7, 2010, by Frank Nutter, President, Reinsurance Association of America.

NYC Office of Emergency Management. 2009. *The 2009 New York City natural hazard mitigation plan.* http://nyc.gov/html/oem/html/about/planning_hazard_mitigation.shtml (accessed November 30, 2010).

NYC Panel on Climate Change. 2010. CLIMATE PROTECTION LEVELS: Incorporating Climate Change into Design and Performance Standards, W Solecki, L Patrick, M Brady, K Grady and A Maroko, authors. *In* Climate change adaptation in New York City: Building a risk management response. *Annals of the New York Academy of Sciences* 1196(1):293-352.

Park J, Cox-Ganser J, Rao C, Kreiss K. 2006. Fungal and endotoxin measurements in dust associated with respiratory symptoms in a water-damaged office building. *Indoor Air* 16:192-203.

Pew Center on Global Climate Change (Pew). 2011a. *Climate action plans.* http://www.pewclimate.org/sites/default/modules/usmap/pdf.php?file=5900 (accessed February 20. 2011).

Pew. 2011b. *U.S. climate policy maps.* http://www.pewclimate.org/what_s_being_done/in_the_states/state_action_maps.cfm (accessed February 20. 2011).

RC/RCCC (Red Cross/Red Crescent Climate Centre). 2011. *About us.* http://www.climatecentre.org/ (accessed April 17, 2011).

RTI International. 2008. *Test/QA plan for mold-resistant building material testing.* Research Triangle Park, North Carolina.

Srebric J. 2010. *Opportunities for green building (GB) grating systems to improve indoor air quality credits and to address changing climatic conditions.* Washington, DC: EPA Indoor Environments Division. http://www.epa.gov/iaq/pdfs/jelena_draft_paper_11-4-10.pdf (accessed February 18, 2011).

Stout DM II, Bradham KD, Egeghy PP, Jones PA, Croghan CW, Ashley PA, Pinzer E, Friedman W, Brinkman MC, Nishioka MG, Cox DC. 2009. American Healthy Homes Survey: A national study of residential pesticides measured from floor wipes. *Environmental Science & Technology* 43(12):4294-4300.

Swiss Re. 2002. *Opportunities and risks of climate change.* http://stephenschneider.stanford.edu/Publications/PDF_Papers/SwissReClimateChange.pdf (accessed February 20, 2011).

Swiss Re. 2010. *Weathering climate change: Insurance solutions for more resilient communities.* http://media.swissre.com/documents/pub_climate_adaption_en.pdf (accessed February 20, 2011).

Thorne PS, Kulhankova K, Yin M, Cohn R, Arbes SJJ, Zeldin DC. 2005. Endotoxin exposure is a risk factor for asthma: The national survey of endotoxin in United States housing. *American Journal of Respiratory and Critical Care Medicine* 172(11):1371-1377.

US Census Bureau. 2008. *American Housing Survey.* http://www.census.gov/hhes/www/housing/ahs/ahs.html (accessed January 11, 2011).

USGCRP (US Global Change Research Program). 2009. *Global climate change impacts in the U.S.* http://downloads.globalchange.gov/usimpacts/pdfs/climate-impacts-report.pdf (accessed February 22, 2011).

USGCRP. 2011. *Program overview.* http://globalchange.gov/about (accessed February 22, 2011).

Visness CM, London SJ, Daniels JL, Kaufman JS, Yeatts KB, Siega-Riz AM, Liu AH, Calatroni A, Zeldin DC. 2009. Association of obesity with IgE levels and allergy symptoms in children and adolescents: Results from the National Health and Nutrition Examination Survey 2005-2006. *Journal of Allergy and Clinical Immunology* 123(5):1163-1169.

WHO (World Health Organization). 2011. *Housing and health.* http://www.euro.who.int/en/what-we-do/health-topics/environmental-health/Housing-and-health/activities/the-large-analysis-and-review-of-european-housing-and-health-status-lares-project (accessed February 8, 2011).

Yan P, Chen H, Zeng D. 2008 Syndromic surveillance stystems—public health and biodefense. *Annual Review of Information Science and Technology* 48:1-96.

4

Air Quality

INTRODUCTION

Indoor air quality (IAQ) is an important component of indoor environmental quality. It has many facets. This chapter focuses on the chemical and particulate pollutants that can be found suspended in air or deposited on or sorbed to indoor surfaces. It specifically addresses organic and inorganic volatile and semivolatile molecular pollutants, and particulate matter. In the case of particles, abiotic materials are emphasized, but there is a brief discussion of allergens associated with pollen and of respiratory health risks associated with algal blooms after floods. IAQ problems associated with moisture and dampness of buildings are addressed in Chapter 5, and biologic IAQ concerns associated with microbial agents, insects and arthropods, and mammals and concerns that arise because of efforts to control them are discussed in Chapter 6.

With regard to the pollutants considered in this chapter, there is little in the published literature that considers together all the key elements in this committee's charge: the effects of climate change on IAQ that would influence public health. However, substantial research has been published on many important components. For example, there is a strong emerging literature on the effects of climate change on outdoor air pollutants (Jacob and Winner, 2009), such as particulate matter (Tagaris et al., 2007) and ozone (Bell et al., 2007; Hogrefe et al., 2004a; Racheria and Adams, 2009), and on related health effects (Kinney, 2008; Tagaris et al., 2009). A voluminous literature characterizes health risks associated with pollutants in outdoor air (Bell et al., 2004; Dockery et al., 1993; Jerrett et al.,

2009; Pope and Dockery, 2006; Pope et al., 2009). Considerable published research documents our understanding of indoor–outdoor relationships of important air pollutants, including particles and ozone (Jia et al., 2008b; Monn, 2001; Wallace, 1996; Weschler et al., 2000). Research has explored the extent to which health risks associated with outdoor pollutants are a consequence of indoor exposures (Weschler, 2006; Wilson and Suh, 1997; Wilson et al., 2000). A large body of work reports on how indoor pollution sources influence IAQ and human health (Jones, 1999; Samet et al., 1987, 1988), including a National Research Council report published three decades ago (NRC, 1981).

The following sections discuss how indoor air pollutant levels might be influenced by climate change. The discussion is organized according to pollutant source category and pollutant class, considering first indoor emission sources and second pollutants of outdoor origin. The treatment is not intended to be comprehensive, but rather broadly illustrative of important IAQ concerns that might be influenced by climate change. Although most of what follows is related to conditions in buildings of the types commonly found in the United States, the chapter concludes with a discussion of an important international public-health problem: exposure to smoke from the indoor combustion of solid biomass and coal in developing countries.

INDOOR SOURCES OF POLLUTANTS

Indoor environments detain pollutants that are emitted indoors. This section reviews important IAQ issues that are associated with indoor pollutant sources and explores how climate change might affect these issues. The emphasis is on conditions in the United States but the discussion is relevant for other countries with similar levels of economic development and similar buildings.

Pollutants from Indoor Combustion

Pollutants released into indoor air cause roughly 100–1,000 times greater human inhalation exposure or dose per unit mass emitted than pollutants released into outdoor air (Smith, 1988). That important observation has been expressed in terms of "intake fraction" (Bennett et al., 2002; Nazaroff, 2008), the ratio of the mass of a pollutant inhaled by an exposed population to the mass of the pollutant emitted from a source. The significance of that point in the present context is that sources have a much larger effect on public health if their pollutants are emitted indoors rather than outdoors. The much higher intake fraction for indoor emissions compared to those outdoors leads to the understanding that small-scale combustion processes that do not burn much fuel can nevertheless raise substantial IAQ concerns and adversely affect public health.

Combustion might be the most important source of air pollution. Indoor combustion for cooking, lighting, and heating has a long and diverse history of contributing to air-pollution exposure. Lopez et al. (2006) ranked "indoor air pollution from [burning] solid fuels" as one of the top 10 leading causes of global mortality and disease. That ranking is based mainly on the use of biomass and coal in rural parts of developing countries. Unvented or incompletely vented combustion also occurs to a substantial extent in developed countries and has demonstrable effects on indoor pollutant concentrations and exposures. Evidence associating those exposures with public-health consequences ranges from suggestive to clear and compelling. Exposures resulting from indoor combustion could be altered in the future in several ways associated with climate change. Influencing factors could include changing prevalence, frequency, or strength of indoor emission rates and also changes in building ventilation conditions.[1] The following paragraphs summarize some of the concerns and provide references to document the nature and importance of the current problems.

Accidental Carbon Monoxide Poisoning

Carbon monoxide (CO) is produced by the incomplete combustion of a carbonaceous fuel. Inhaled CO forms carboxyhemoglobin in the blood, whose presence interferes with transport and delivery of oxygen to tissues and organs. Excessive acute exposures result in illness or death. Chronic lower-level exposures may also have health consequences, but the available empirical evidence is weaker than that for acute poisonings.

CO is regulated as a pollutant in ambient air. Mainly through strong improvements in automotive emission-control technology, urban air CO levels have become well controlled, and almost every area of the United States meets the National Ambient Air Quality Standard for CO (EPA, 2010b).

Despite improvement in outdoor levels, CO remains an important air pollutant. Over the past few decades, hundreds of accidental and fatal acute CO poisonings have occurred each year in the United States (Cobb and Etzel, 1991; King and Bailey, 2008; Mott et al., 2002). The incidence has declined substantially. One important factor is improvements in the control of motor-vehicle emissions. Mott et al. analyzed CO-associated mortality statistics and concluded that, "if rates of unintentional CO-related deaths had remained at pre-1975 levels, an estimated additional 11,700 motor-vehicle-related CO poisoning deaths might have occurred by

[1] Building tightening and reduced ventilation rates are further discussed in Chapter 8.

1998." Holmes and Russell (2004) remarked that the reduction in accidental deaths resulting from improvements in motor-vehicle emission controls "is not accounted for in EPA's [the Environmental Protection Agency's] recent reports on the benefits and costs of the [Clean Air Act], yet it dwarfs the estimated direct benefits ascribed to CO control." In a detailed study of CO poisoning deaths in California during the period 1978–1988, Girman et al. (1998) found that alcohol was a factor in 31% of the cases and that important combustion sources other than motor vehicles included heating or cooking appliances, charcoal grills and hibachis, small engines, and camping equipment. An assessment for Florida over the period 1999–2007 revealed that accidental CO poisonings "were primarily due to motor vehicle exhaust (21%–69%) and generator exposure (12%–33%), and the majority (50%–70%) occurred within the home" (Harduar-Morano and Watkins, 2011).

In the context of climate change, a particular concern about CO exposure arises from the use of emergency electricity generators that burn liquid fuels, such as gasoline. The use and reliability of centrally generated power might be degraded because of climate change for several reasons. For example, hotter summer afternoons may lead to more intense use of air conditioners and thus increase the frequency of service-demand overloads that cause brownouts and blackouts. Severe storms can also cause electricity service disruption. In such cases, people may rely more heavily on their own electricity generators. If the generators are used indoors, or even outdoors but too close to an indoor environment, unhealthful CO exposures can result. Increases in emergency-room and other hospital visits caused by CO poisoning have been reported in association with power outages (Muscatiello et al., 2010), major storms (Van Sickle et al., 2007), and floods (Daley et al., 2001).

A staff report from the Consumer Product Safety Commission (Hnatov et al., 2009) indicated that in 2005 an estimated 27 generator-related CO fatalities were associated with five hurricanes (Katrina, Rita, Wilma, Dennis, and Isabelle). And an estimated 21 generator-related CO fatalities were associated with ice storms, including major storms in the midwestern United States in January and in the Carolinas and Georgia in December.

In addition to electricity generators, shifts in fuel-use patterns during power outages may contribute to increased indoor CO levels. Of concern would be the use of natural-gas–fueled and petroleum-fueled stoves for heating, excessive reliance on unvented combustion-based space heaters, and use of charcoal briquettes or wood stoves indoors for cooking (Hampson and Stock, 2006, Hnatov, 2009).

One expects there to be many more poisonings that result in illness than in death. Analyses of the demand for poison control center services reveal a pattern similar to that in emergency rooms. Klein et al. (2007)

noted a nearly 50% increase in suspected CO poisoning calls in the days after a widespread blackout on the East Coast of the United States in 2003, and Forrester (2009) found more such calls in the counties that were in the disaster area declared for Hurricane Ike than in other counties in Texas. It is reasonable to believe that the prevalence of CO-induced illness is larger than that recorded in the emergency-room statistics because illnesses that are not considered severe might not be reported. A recent study evaluating the use of a web-based query system for public health surveillance reported almost 25,000 CO-related hospitalizations across the United States in 2005, of which approximately 4,200 were confirmed CO-related poisonings (Iqbal et al., 2010). These data were intended to exclude intentional and fire-related CO exposures.

Other factors may also contribute to increased public health risks associated with indoor CO exposures. For example, the Department of Housing and Urban Development's 2009 American Housing Survey found that just 36% of homes nationwide reported having a working CO detector.[2] People of lower socioeconomic status may be more likely to use stoves or unvented space heaters as a heat source (CDC, 1997) and less likely to have working CO detectors (Runyan et al., 2005). Some groups may hold mistaken beliefs about CO. For example, a survey conducted among residents of low socioeconomic status in northern Mexico by Galada et al. (2009) found that a large majority of respondents mistakenly believed that CO could be detected by sight or smell.

Cooking

Cooking causes air-pollutant exposures that have potential public-health significance. The most severe problems occur from burning of solid biomass fuels or coal, especially in unimproved stoves, in the rural parts of developing countries. The relationship of those concerns to climate change is discussed toward the end of this chapter. However, even when relatively clean fuels are used for cooking in developed countries, indoor air-pollutant exposures with potential public-health consequences can arise. For example, the use of natural gas as a cooking fuel is associated with increased indoor exposures to nitrogen dioxide (NO_2), a byproduct of the combustion process (Marbury et al., 1988; Spengler et al., 1994). In a study in the United Kingdom, the use of gas cooking appliances, rather than electric, was associated with respiratory morbidity in women (but not men, possibly women had higher exposure) (Jarvis et al., 1996). Exposure of children to higher indoor NO_2 levels has also been reported to be associated with re-

[2] As of January 2010, 25 states—including Florida, Texas, and California—required some or all residences to have CO detectors (National Conference of State Legislatures, 2010).

spiratory symptoms (such as wheeze) but not pulmonary function (Neas et al., 1991). In a population of infants at risk for asthma, "the frequency of reported respiratory symptoms in the first year of life was associated with NO_2 levels not currently considered to be harmful" (van Strien et al., 2004). However, another study did not find an association between NO_2 level and respiratory illnesses in infants (Samet et al., 1993). A study of asthmatic children in inner-city environments found that indoor NO_2 levels were substantially elevated in homes with gas stoves and that "higher levels of indoor NO_2 are associated with increased asthma symptoms in nonatopic children and decreased peak flows" (Kattan et al., 2007). Early life exposure to household gas appliances has also been associated with negative neuropsychological development (Morales et al., 2009). Valero et al. (2009) investigated the determinants of exposure for a cohort of Spanish women and found that personal NO_2 levels were "strongly influenced by indoor NO_2 concentrations." They also found that outdoor NO_2 levels and the use of gas appliances were important determinants of indoor NO_2 levels, whereas no significant association "was found between personal or indoor NO_2 levels and exposure to environmental tobacco smoke (ETS) at home."

Cooking can also substantially increase indoor fine-particle mass concentrations ($PM_{2.5}$) (Abt et al., 2000; Buonanno et al., 2009; Evans et al., 2008; Olson and Burke, 2006; Wallace et al., 2004). Fumes from Chinese-style cooking with hot oil have been shown to be mutagenic (Chiang et al., 1997), and this cooking style has also been reported to be a risk factor for lung cancer in nonsmoking women in Taiwan (Ko et al., 1997). Exposure to ultrafine particles can be substantially increased by emissions from cooking (Bhangar et al., 2011; Mullen et al., 2010). Emissions of ultrafine particles can be caused not only by the combustion of cooking fuel but from high temperatures associated with electric cooking elements (Wallace et al., 2008).

Climate change could affect the indoor concentrations of cooking-associated pollutants in the United States and other developed countries in several ways. First, it may be that a mitigation response to climate change drives a movement toward smaller per-capita housing space (with lower life-cycle environmental effects) and with lower air-exchange rates (to save heating and cooling energy). If so, emissions from cooking would be diluted into a smaller volume and would persist for longer times, and these changes would tend to increase concentrations and exposures associated with a given level of cooking. Second, climate-change mitigation goals might push cooking away from the use of natural gas and toward a heavier reliance on electricity (assuming that electricity would be generated from lower-carbon sources than today). Such a shift would reduce associated exposures to NO_2 and to the ultrafine particles formed in combustion flames. Third, tighter building envelopes resulting from weatherization efforts might reduce the efficacy of local exhaust hoods and fans for removing cooking-related emissions before they enter indoor air. Dampers have been developed that automatically open when exhaust fans are activated to permit additional ventilation supply air to flow freely into a building, thereby mitigating this otherwise adverse effect of weatherization.

Space Heating

In the United States, combustion for space heating can sometimes be associated with substantial pollutant emissions, especially because of the relatively large amounts of fuel used for home heating compared with, for example, cooking. When on-site combustion is used to generate heat, it is usually the case that the heat is first extracted from the combustion gases and then the byproducts are vented to the outside. Leakage may occur, and some of the generated pollutants can enter the occupied indoor space of the same building for which the heat is being generated. In addition, combustion for heating is sometimes unvented by design, in which case all the byproducts formed are emitted into the indoor environment with the generated heat. The direct evidence that links household heating with health effects is sparse. Household use of kerosene heaters and fireplaces for heating was found to be associated with respiratory symptoms in nonsmoking women in Connecticut and Virginia during the 1990s (Triche et al., 2005). A study of coroners' reports in California found that unvented combustion heating appliances and cooking indoors with charcoal were associated with CO deaths (Liu et al., 2000).

Climate change could induce several shifts that would affect indoor air-pollutant exposures associated with heating. First, if average temperatures rise, as is expected, less heating may be needed, and—other things being equal—there would tend to be less associated pollution exposure. Climate-change mitigation efforts may lead to better insulation of buildings, which also would lessen heating requirements. Second, there could be shifts in the types of heating sources used. Mitigation efforts could serve as a driving force for substituting electricity (from low-carbon sources) for fossil-fuel combustion—a change that would tend to improve IAQ. In contrast, mitigation goals might also encourage greater use of wood as a household heating fuel. Wood contains contemporary rather than fossil carbon. If grown and harvested sustainably and if burned completely, wood combustion could have little or no net climate impact. However, as practiced today, residential wood combustion is associated with degraded neighborhood air quality owing to emissions exhausted from chimneys and is associated with degraded IAQ in the households that burn the wood owing to leakage of combustion byproducts into the indoor environment (Gustafson et al., 2008; Traynor et al., 1987). If done poorly, increased wood-based heating could exacerbate IAQ problems associated with residential wood combustion.

Another trend that might emerge and that would tend to degrade IAQ is greater reliance on unvented combustion-based space heaters. Devices of this type have a high thermal efficiency because all the generated heat is discharged indoors. However, their use can cause substantially increased indoor concentrations of NO_2, sulfur dioxide (SO_2), and particulate matter (Francisco et al., 2010; Leaderer, 1982; Leaderer et al., 1990; Ruiz et al., 2010; Wallace and Ott, 2011).

An additional concern associated with climate change and home heating is building envelope tightness. Efforts to save energy by reducing the leakiness of building envelopes can increase the risk of "backdrafting," in which air flows into a building through the exhaust flue, instead of flowing out of the building, and carries combustion byproducts with it. The causes and consequences of backdrafting have received some attention in the literature (Nagda et al., 1996), but the prevalence even in current conditions in the building stock has not been well characterized, and it is not clear what to expect in this regard as a consequence of climate change.

Smoking

Habitual indoor smoking adversely affects IAQ and public health. Sidestream smoke (from the smoldering tobacco product) and exhaled mainstream smoke together constitute the source of environmental tobacco smoke (ETS). Smoking indoors has a strong influence on indoor levels of $PM_{2.5}$ (Hyland et al., 2008; Nazaroff and Klepeis, 2004). ETS is also an important cause of environmental exposure to some hazardous air pollutants, including acrylonitrile, 1,3-butadiene, acetaldehyde, acrolein, and formaldehyde (Nazaroff and Singer, 2004). Evidence indicates that several severe adverse health effects are associated with ETS exposure, including acute myocardial infarction (Lightwood and Glantz, 2009), lung cancer (Fontham et al., 1994), and a host of respiratory health problems in children (DiFranza et al., 2004). Over the past few decades, there has been a marked reduction in exposure to ETS in the US population, as reflected in lower concentrations of serum cotinine in nonsmokers (Pirkle et al., 2006). The decline is a consequence mainly of declines in the amount of smoking that occurs indoors rather than of changes in the building stock.

In a future influenced by climate change, exposure of nonsmokers to ETS will be determined to a great degree by the prevalence and intensity of smoking in indoor spaces. In the United States, smoking in public places has become uncommon. However, smoking in private residences continues: Singh et al. (2010) estimated that 7.6% of children in the United States are exposed to ETS in their own homes. Exposures to ETS occur not only in the residence in which smoking occurs but, in the case of multifamily dwellings, in neighboring units (Bohac et al., 2011). Some parts of the US population have a relatively high prevalence of indoor smoking. For example, a study of 100 asthmatic children in inner-city Baltimore revealed an indoor smoking prevalence of 46% and found that average indoor $PM_{2.5}$ and PM_{10} levels were 33–54 μg/m^3 higher in smoking than in nonsmoking households (Breysse et al., 2005). In another study, fine-particle concentrations were sampled over two-week periods in 294 inner-city homes with asthmatic children (Wallace et al., 2003). In these homes, the average particle mass concentration, 27.7 μg/m^3, was considerably higher than the average concurrently measured outdoor concentration, 13.6 μg/m^3. Smoking occurred in 101 of the homes (34%) and caused an average increase of 37 μg/m^3 for indoor fine particle levels. Other identified sources—frying, smoky cooking events, and use of incense—made smaller contributions, 3–6 μg/m^3.

It is unknown how smoking patterns that would affect indoor ETS will evolve. In particular, it is not clear that indoor smoking behaviors would be influenced by climate change. Changes in tobacco or in tobacco products could alter the ETS characteristics associated with indoor smoking, and there is some published evidence that tobacco itself might be altered in response to changing temperature and atmospheric CO_2 levels (Ziska et al., 2005).

Changes in the residential building stock that are a consequence of climate-change concerns could influence exposure to ETS. Currently, unintended airflow pathways in multiunit residential buildings can lead to exposures to secondhand smoke in the units of nonsmokers (Kraev et al., 2009; Wilson et al., 2011; Winickoff et al., 2010). Mitigation measures to reduce energy use in buildings could lead to systematically lower ventilation rates and alteration of internal airflows that could cause higher concentrations and exposures to secondhand smoke. For a given characteristic, such as number of cigarettes smoked indoors per day, any of those changes would tend to increase exposures to ETS indoors.

Candles, Incense, and Other Small-Scale Combustion Processes

Pagels et al. (2009) summarize some of the IAQ concerns related to indoor candle use. The local high temperature created by a candle flame can volatilize candle components that are then emitted to indoor air. Some candles have metal-cored wicks that emit lead at rates sufficient to pose health concerns (Wasson et al., 2002). Depending on combustion conditions, the candle flame also produces soot particles and other products of incomplete combustion that are emitted indoors (Fine et al., 1999).

According to the National Candle Association (National Candle Association, 2011), US retail sales of candles are roughly $2 billion per year, and "candles are used in 7 out of 10 US households." Given the type and scale of emissions summarized in the previous paragraph, the potential for air-pollutant exposure due to candle use would seem to be substantial, but scientific data that would permit one to quantify the extent of indoor use and the resulting air-pollutant exposures are lacking.

In developing countries, combustion-based technologies, such as candles and kerosene lamps, are commonly used to provide lighting. Those are inherently inefficient in converting chemical energy into light (Mills, 2005). The air-pollutant exposure consequences of combustion-based lighting are expected to be substantial but have only begun to be explored (Apple et al., 2010).

Indoor air-pollutant emissions from other small-scale combustion sources have been investigated, and a few illustrative examples are noted here. Jetter et al. (2002) studied the emissions from burning incense and concluded that "incense smoke can pose a health risk to people due to inhalation exposure of particulate matter." Liu et al. (2003) characterized emissions and IAQ effects of burning mosquito coils, which are commonly used in households in Asia, Africa, and South America. They concluded that "exposure to the smoke of mosquito coils similar to the tested ones can pose significant acute and chronic health risks."

As in the case of other indoor combustion activities, climate change would affect IAQ and potentially public health if it were accompanied by a change in the source emission rate (for example, owing to a change in use) or were accompanied by a change in the other factors that influence exposures associated with a given magnitude of emissions. There is no good basis of expectations of use patterns of small-scale combustion sources. As noted in connection with other combustion sources, reduced household volume per occupant and lower air-exchange rates might be consequences of efforts to mitigate anthropogenic effects on climate, and such changes would tend to increase air-pollutant exposures that result from indoor combustion sources.

Radon and Its Decay Products

Indoor radon is a major cause of the public's health-relevant radiation exposure. Exposure to increased residential radon is an important risk factor for lung cancer. On the basis of a combined analysis of 13 studies that collectively involved 7,148 lung-cancer cases and 14,208 controls, Darby et al. (2005) concluded that residential radon is "responsible for about 2% of all deaths from cancer in Europe." In a parallel North American effort encompassing 7 studies that collectively assessed 3,662 cases and 4,966 controls, Krewski et al. (2005) reported that their results "provide direct evidence of an association between residential radon and lung cancer risk, a finding predicted using miner data and consistent with results from animal and in vitro studies."

Radon-222 (radon), the most health-significant of the three naturally occurring isotopes, is generated by the radioactive decay of radium-226, a ubiquitous trace element in the earth's crust. Being an inert gas, radon has the potential to migrate from its parent material during its short lifetime (half-life, 3.8 days) and enter indoor or outdoor air, where humans may encounter it. Radon does not directly pose a substantial health hazard. However, its radioactive decay marks the beginning of a sequence of short-lived products. Those radon decay products—isotopes of bismuth, lead, and polonium—are chemically reactive and, when inhaled, can be retained on respiratory tract tissues; later radioactive decays irradiate lung cells. Of particular health concern are the alpha-particle emissions from the decay of polonium-218 and polonium-214. It is the radiation damage caused by those alpha-particle emissions that creates the lung-cancer risk associated with exposure to residential radon. The epidemiologic evidence is consistent with a linear no-threshold dose–response model. Health risks posed by a given level of radon exposure are much higher in smokers than in nonsmokers (Ginevan and Mills, 1986).

The three main sources of indoor radon are soil near a building's foundation; earthen building materials, such as concrete; and tap water from underground sources. In aggregate for the entire building stock, soil is the most important radon source, although the other two sources dominate in some buildings. The significance of soil as a source of indoor radon depends on the radium content of the soil, on the permeability of the soil, and on the degree of coupling between the indoor space of the building and the pore air in the underlying and adjacent soil (Nazaroff, 1992). The only important mechanism for removing radon from indoor air is ventilation. However, the effective radiation dose to lung tissue associated with a given level of indoor radon depends on the dynamic behavior of the short-lived decay products (Porstendörfer, 1994), which can be influenced not only by the ventilation rate but by such factors as indoor particle levels, active air filtration, and the intensity of indoor air movement.

Annual average residential radon levels in the United States have been estimated to have an arithmetic mean of 46 ± 4 Bq/m^3 (1.25 ± 0.12 pCi/L) with an estimated 6% of dwellings exceeding the EPA mitigation level of 148 Bq/m^3 (4 pCi/L) (Marcinowski et al., 1994). EPA has estimated that 20,000 US lung-cancer deaths a year are radon-related (Pawel and Puskin, 2004). Radon-control systems are well established in principle for maintaining low indoor radon concentrations (Rahman and Tracy, 2009). However, challenges remain to identify buildings with high concentrations and to apply effective controls, where appropriate, in both existing and new buildings.

Climate change might induce shifts in indoor radon and decay-product concentrations for several reasons, although the direction and scale of the changes are difficult to predict. Changes that would reduce ventilation rates would tend to increase indoor radon levels and might also alter the effective radiation dose received. Constructing buildings with or near materials that have high radium content should be avoided irrespective of climate-change concerns. The goal of improving the energy performance of buildings might induce increased use of subterranean spaces for habitation or stronger thermal coupling of building interiors to climate-buffered underground zones. Care would be needed in such cases to prevent radon levels from increasing in the occupied spaces.

Volatile Organic Compounds and Semivolatile Organic Compounds

Organic compounds constitute a diverse set of chemicals that have a broad array of properties. For the discussion here, organic compounds are divided into two primary groups. Volatile organic compounds (VOCs) are species that have high enough vapor pressures to volatilize substantially and, when unconfined, to be found predominantly in the gas phase. Semivolatile organic compounds (SVOCs) are preferentially found in the condensed phase, but they still have sufficient volatility to be present in the vapor-phase. For SVOCs, the saturation vapor pressure is roughly in the range 10^{-9} to 10 Pa; VOCs have vapor pressures higher than 10 Pa. Organic compounds that have extremely low volatility can also be present purely in the condensed phase and could still contribute to IAQ concerns as constituents of particulate matter. An important example of this category would be polycyclic aromatic hydrocarbons (PAHs) that have many rings.

Excessive exposures to volatile and semivolatile organic compounds indoors raise a broad range of public health concerns. For example, many organic chemicals that have been classified by the US federal government as hazardous air pollutants (HAPs) are present at significant levels indoors. This classification applies to pollutants that "cause or may cause cancer or other serious health effects, such as reproductive effects or birth defects, or adverse environmental or ecological effects (EPA, 2009b). Some of these chemicals may be used as constituents of construction and finishing materials, or as ingredients in consumer products used indoors. Seminal research from the 1980s known as the Total Exposure Assessment Methodology (TEAM) studies showed that concentrations in personal air (heavily influenced by indoor conditions, because people spend most of their time indoors) commonly exceeded outdoor levels for several toxic air pollutants, including benzene, carbon tetrachloride, chloroform, dichlorobenzene, tetrachloroethylene, 1,1,1-trichloroethane, trichloroethylene, and xylenes (Wallace et al., 1985, 1987, 1988).

In addition to concerns related to their status as toxic air pollutants, exposure to VOCs has also been associated with such health effects as allergic symptoms, asthma, and symptoms of sick-building syndrome (Garrett et al., 1999; Norbäck et al., 2000; Smedje et al., 1997; Ten Brinke et al., 1998). Establishing definitive links between exposure to those compounds and these types of health effects is challenging because the amount of exposure sustained by study subjects and the conditions under which they are exposed generally are beyond the direct control of the investigator. In addition, human populations are routinely exposed to multiple contaminants whose individual, let alone joint, effects are not known (Cohen and Gordis, 1993).

Research on indoor VOCs began in the 1980s. Research on indoor SVOCs is much less developed. However, many studies of subclasses of SVOCs indoors—including pesticides, polybrominated diphenyl ethers (flame retardants), and phthalates (plasticizers)—have been published. Several review articles have been published on the occurrence and potential health significance of VOCs and SVOCs indoors (Brown et al., 1994; Jones, 1999; Logue et al., 2011; Mendell, 2007; Rudel and Perovich, 2009; Salthammer and Bahadir, 2009; Weschler and Nazaroff, 2008; Wolkoff, 1995; Wolkoff et al., 1997). Logue et al. (2011) compared published concentration data on VOCs in US houses with health-based exposure guidelines and standards. They identified seven organic compounds as "priority hazards based on the robustness of measured concentration data and the fraction of residences that appear to be impacted." In alphabetical order, the seven are acetaldehyde, acrolein, benzene, 1,3-butadiene, 1,4-dichlorobenzene, formaldehyde, and naphthalene. An important attribute of SVOCs is their potentially long persistence indoors. Available evidence suggests that these may pose legacy pollution concerns (Weschler and Nazaroff, 2008).

Focusing on sensory irritation and other perceived IAQ effects, Wolkoff et al. (1997) have called attention to the importance of secondary pollutant formation indoors due to reactions involving organic compounds and oxidizing agents, such as ozone (O_3) and NO_2. Mendell (2007) reviewed 21 studies from the "epidemiologic literature on associations between indoor residential chemical emissions, or emission-related materials or activities, and respiratory health or allergy in infants or children." He found that the most frequently identified risk factors included "formaldehyde or particleboard, phthalates or plastic materials, and recent painting."

Emissions of VOCs indoors tend to be higher after new construction and renovation activities because of releases from finite-capacity reservoirs in wood-based products, paints, floor finishes, glues, and other construction and finishing materials (Dales et al., 2008; Herbarth and Matysik, 2010). House dust is an important repository for SVOCs and other particle-bound contaminants (Butte and Heinzow, 2002). Results of studies of house dust have demonstrated the presence of polychlorinated biphenyls (PCBs), PAHs, plasticizers (phthalates and phenols), flame retardants, other organic xenobiotics, and inorganic constituents (Weschler and Nazaroff, 2010). Dust ingestion can also be an exposure pathway of concern for SVOCs (Roosens et al., 2009). Infants are generally affected more by dust ingestion than adults because of their contact with floors and their high level of hand-to-mouth activity.

In subsections that follow, indoor exposure conditions and associated health concerns are summarized for two important examples in the broader category of VOCs and SVOCs: formaldehyde and endocrine-disrupting chemicals. Overall, however, the state of knowledge about VOCs and SVOCs in indoor environments and their consequences for public health is far from complete. The chemicals are diverse in their characteristics and complex in their dynamic behavior. Conditions vary with time in any given building and can also vary markedly among buildings. Concentrations are influenced by a variety of factors, some of which reflect properties of the chemicals, some of which depend on properties of the buildings into which they are emitted, and some of which depend on actions of building occupants, for example, in relation to frequency and intensity of use of products that contain the chemical of concern. Occupant behaviors also affect how concentrations are related to exposures.

Increasingly, biomarkers are being used to measure body burdens of environmental chemicals or chemical byproducts in human tissue (Paustenbach and Galbraith, 2006; Sexton et al., 2006). Some recent work has focused on prenatal life and infancy as highly vulnerable periods of development. Investigations have monitored indoor exposures to multiple chemicals and birth outcomes (Eskenazi et al., 1999; Herbstman et al., 2010; Perera et al., 2003; Rosas and Eskenazi, 2008). However, there is still an inadequate understanding of the relationship between environmental concentrations of VOCs and SVOCs, resulting biomarker levels, and associated health outcomes.

Available information suggests that concerns about the influence of climate change on exposures and public-health risks associated with VOCs and SVOCs are substantial enough to warrant further attention, but it is insufficient to support substantive conclusions. In his review of indoor pollutants over the past 50 years, Weschler (2009) made an important point in this regard, stating that

> Many of the chemicals presently found in indoor environments, as well as in the blood and urine of occupants, were not present 50 years ago. Given the public's exposure to such species, there would be exceptional value in monitoring networks that provided cross-sectional and longitudinal information regarding pollutants found in representative buildings.

Formaldehyde

Formaldehyde is a chemical of concern as a carcinogen and as an airway irritant. The International Agency for Research on Cancer has classified formaldehyde as "carcinogenic to humans" (IARC, 2006). Formaldehyde is listed by EPA as a hazardous air pollutant (EPA, 2008) and as a toxic air contaminant in the state of California (OEHHA, 2007). Indoor formaldehyde levels can be elevated because of emissions from indoor sources, such as wood-based products, paints, O_3-initiated chemical reactions, and combustion (Salthammer et al., 2010). Acute effects, such as eye, nose, and throat irritation have been observed after controlled exposures at levels of at least 1,230 µg/m^3 (Kulle, 1993); however, persons who have asthma or allergic sensitization have been shown to respond with bronchial symptoms to exposure at concentrations as low as 100 µg/m^3 (Casset et al., 2006). An Institute of Medicine report on IAQ and asthma concluded that there was limited or suggestive evidence of an association between formaldehyde exposure and wheezing and other respiratory symptoms (IOM, 2000). It also concluded that there was inadequate or insufficient evidence to determine whether an association between formaldehyde exposure and asthma exists. At moderate to high exposure, formaldehyde can be irritating, producing such symptoms as sore throat, cough, scratchy eyes, and nosebleed. Some people are more sensitive than others, and an exposure that causes no problems for some people can make other people sick or uncomfortable.

There is no US federal regulation or standard for formaldehyde levels in residential settings. In 2008, however, California established a regulation limiting the emissions of formaldehyde from composite wood products[3] that are common indoor sources of formaldehyde (California Environmental Protection Agency, 2011). Also, the California Air Resources Board has published a guideline value of 27 ppb (33 µg/m^3) for formaldehyde in residences to avoid irritant effects (Cal/EPA ARB, 2004).

Residential formaldehyde levels are influenced by ventilation rates. In a sample of 122 new homes in California, formaldehyde levels were inversely correlated to the air-exchange rate (Offermann, 2009). Based on 24-h average samples in 105 of these homes, more than 90% exceeded California's chronic reference exposure level of 9 µg/m^3 (OEHHA, 2007) and 59% exceeded the 33 µg/m^3 indoor guideline value. A study of 96 homes in Québec City found that higher concentrations of formaldehyde were associated with lower air-exchange rates (Gilbert et al., 2006). A monitoring program that sampled formaldehyde in 252 homes in Germany also found that indoor levels were inversely correlated with air-exchange rates (Salthammer

---

[3] California Code of Regulations §93120. *Airborne Toxic Control Measure to Reduce Formaldehyde Emissions from Composite Wood Products.*

et al., 1995). That study also found that indoor formaldehyde levels were positively, albeit weakly correlated with higher indoor temperatures and relative humidities. Higher indoor temperatures and relative humidities might be expected in some indoor environments as a consequence of climate change.

In response to the first modern energy crisis in the 1970s, urea-formaldehyde foam insulation (UFFI) was widely used as a retrofit building insulation material. Emissions of formaldehyde, perhaps exacerbated because of improper or inappropriate use in some cases, led to a concern among Canadian authorities, where the use of UFFI was banned in 1980 (CMHC, 2011). In 1982, UFFI was banned by the US Consumer Products Safety Commission (CPSC, 1982); however, that ban was overturned by a legal ruling in 1983 (CPSC, 1983). Research conducted on homes with UFFI in Toronto only showed a moderate increase in formaldehyde levels (median = 38 ppb, N = 571) compared with untreated homes (median = 31 ppb; N = 231) (Broder et al., 1988). Remedial intervention to remove UFFI from some homes showed post-removal improvement in many health status indicators, although the "improvement in health status among the UFFI removal subset was not associated with any significant diminution of formaldehyde exposures" (Broder et al., 1991). L'Abbé and Hoey (1984) reviewed the evidence available at that time and concluded that "epidemiologic studies have not established causation or an association between UFFI exposure and health effects."

After Hurricane Katrina (in 2005), hundreds of temporary housing trailers provided by the Federal Emergency Management Agency were found to have elevated levels of formaldehyde (Maddalena et al., 2009). After learning of the potential concern regarding the air quality in the trailers used for temporary housing after hurricanes, the Centers for Disease Control and Prevention stated that residents living in temporary trailer housing should open windows as much as possible to let in fresh air, keep indoor temperatures at the lowest comfortable setting, run an air conditioner or dehumidifier to control mold, and spend as much time outdoors as possible. Children and elderly people and those with chronic diseases, such as asthma, were particularly encouraged to spend time outside. However, in the event of unhealthful outdoor temperatures and high outdoor levels of pollutants, vulnerable populations may be forced to choose between unhealthful indoor exposures to formaldehyde and other pollutants from indoor sources and exposures to heat and outdoor air pollution. Persons in buildings that lack adequate ventilation and air conditioning would be particularly vulnerable.

Endocrine-Disrupting Chemicals

Many of the chemicals classified as SVOCs and used in products found indoors have demonstrated or suspected endocrine-disrupting properties (Rudel and Perovich, 2009). Human exposure to indoor air and to dust enriched with endocrine-disrupting chemicals released from indoor sources has become an issue of increasing concern (Hwang et al., 2008; Rudel et al., 2003, 2010). Semivolatile, endocrine-disrupting chemicals of concern include brominated organics, such as polybrominated diphenyl ethers (PBDEs), used as flame retardants; phthalates, used as plasticizers; PCBs, historically used in caulks and many other products; and pesticides, including organochlorines (such as DDT), organophosphates (such as chlorpyrifos), and pyrethroids (such as permethrin). The following paragraphs provide brief summaries of the concerns for the first three of these classes; pesticides are discussed in Chapter 6.

Epidemiologic studies have demonstrated some associations between indoor levels of SVOCs and adverse health effects, such as childhood leukemia, neurologic disorders, non-Hodgkin lymphoma, and respiratory symptoms (Bornehag et al., 2004; Butte, 2004; Colt et al., 2006). Studies have also produced suggestive evidence that prenatal exposure to those substances may have a deleterious effect on neurodevelopment (Chevrier et al., 2008; Eskenazi et al., 2006; Jacobson and Jacobson, 1996; Jacobson et al., 1990; Rogan and Gladen, 1991; Rogan et al., 1986).

PBDEs, a major class of flame retardants, are ubiquitous environmental contaminants with particularly high concentrations in humans in the United States (Fischer et al., 2006). For the purpose of retarding ignition, these chemicals are added to consumer electronic cases and other materials used indoors. The compounds are slowly released from the products during their life cycles (Alcock et al., 2003). Biomonitoring has shown children's levels to be 2–5 times higher than those of their parents, perhaps because of children's greater exposure to and ingestion of house dust into which PBDEs preferentially partition (Fischer et al., 2006). The potential health hazards of PBDEs are attracting increasing scrutiny. They have been shown to reduce fertility in humans at levels found in households (Harley et al., 2010). Children who have higher concentrations of PBDEs in their umbilical-cord blood at birth have been found to score lower on neurodevelopment tests at the ages of 1 and 6 years (Herbstman et al., 2010). Although the pathways through which PBDEs get into people are not fully understood, releases from indoor construction and furnishing materials, aging and wear of consumer products, and direct exposure during use (for example, from furniture) are potentially important contributors.

Phthalates are one of the more abundant contaminants in household dust. At one time, they were considered safe, but studies have revealed that they may pose endocrine-disruption risks and exhibit reproductive and developmental toxicity (ATSDR, 2002; Duty et al., 2003; Hauser and Calafat, 2005). EPA announced that a review of the safety of these chemicals would begin in fall 2010—an indication of concern about phthalates based on toxicity (particularly in the development of the male reproductive system), prevalence in the environment, widespread use, and resulting human exposure (EPA, 2010a). Adverse effects on the development of the male reproductive system may be the most sensitive health outcomes of phthalate exposure according to studies of laboratory animals. Several studies have shown associations between phthalate exposures and human health, although no causal link in humans has been established. Biomonitoring data from 1999–2000 and 2001–2002 demonstrated that children have the highest exposures to phthalates of all groups monitored; other biomonitoring data have shown in utero exposures to phthalates (CDC, 2005). Indoor environmental exposures may be important contributors to total uptake (Wormuth et al., 2006).

Although the use of PCBs has been banned or restricted for decades, they are still being found indoors in older buildings at levels that are considered to be of concern for human health. Potential health risks posed by PCBs remain high in some indoor environments because of weak removal processes and long-term release from sources (Herrick et al., 2004; Rudel et al., 2008).

Carbon Dioxide

Indoor exposures to CO_2 are likely to increase as a consequence of climate change. The atmospheric background concentration of CO_2 is rising. The preindustrial level was approximately 280 ppm; the level is about 390 ppm now and continues to rise by a few parts per million per year (Keeling, 2009; NRC, 2010). There is also evidence of a rural–urban gradient—levels of CO_2 in outdoor air are higher in urban than in rural environments (George et al., 2007)—and the percentage of people living in urbanized areas is increasing. In the United States in 2000, 58% of the population lived in an urbanized area with a population above 200,000 (US Department of Transportation, 2004).

CO_2 levels are substantially higher in occupied buildings than outdoors. Unvented combustion sources can contribute, but the main indoor source of CO_2 is the exhaled breath of building occupants. The metabolic production of CO_2 by humans depends on diet and activity level. A sedentary adult typically generates CO_2 at 0.31 L/min (ASHRAE, 2010), corresponding to 34 g/h (at an atmospheric pressure of 1 atm and a temperature of 293 K).

CO_2 levels are commonly used to guide ventilation practice in occupied buildings. In that case, CO_2 is serving as a marker of human bioeffluents.

Research shows that "maintaining a steady-state CO_2 concentration in a space no greater than about 700 ppm above outdoor air levels will indicate that a substantial majority of visitors entering a space will be satisfied with respect to human bioeffluents (body odor)" (ASHRAE, 2010).

In common practice, building ventilation requires energy use. In mechanically ventilated buildings, fan power is required to move air through ducts. When the temperature or humidity of the outdoor air is not suitable for establishing desired indoor conditions, energy is used to condition the ventilation air. The desire to mitigate effects of climate change is creating pressure to reduce ventilation rates in buildings and the use of energy. Hence, one might reasonably expect indoor CO_2 to rise in a climate-change–influenced future for two reasons: increased baseline levels due to rising outdoor CO_2 levels, especially in cities; and reduced ventilation rates in buildings as part of a mitigation strategy.

CO_2 is an acid gas. At high levels, "inhalation of CO_2 can produce physiological effects on the central nervous, respiratory, and the cardiovascular systems" (US Department of Labor, 1990). Recognizing its potential for frank adverse health effects, the Occupational Safety and Health Administration maintains an occupational standard for CO_2, with a "transitional limit" of 5,000 ppm for the 8-h time-weighted average (US Department of Labor, 1989).

Occupational standards for pollutants are typically set at much higher levels than would be appropriate for the general public. In the United States, there are no health-based guidelines or standards for CO_2 itself that would apply for the general public in all indoor environments. In Germany, a governmental work group recommended that "based on health and hygiene considerations: concentrations of indoor air carbon dioxide levels below 1000 ppm are regarded as harmless, those between 1000 and 2000 ppm as elevated and those above 2000 ppm as unacceptable" (Ad-hoc Work Group, 2008; translated in Heinzow and Sagunski, 2009).

Existing literature does not provide a clear answer to the question of whether public-health consequences are associated with exposure to CO_2 indoors at the levels at which they might occur in a future influenced by climate change, but it does contain some important clues. A study by Bekö et al. (2010) measured CO_2 in the bedrooms of 500 Danish children to characterize ventilation conditions during sleep. They found that 6% of the rooms had levels over 3,000 ppm (20-min running mean) at some time during the night, which is well above that viewed as "unacceptable" by the German indoor-air working group cited above. Shendell et al. (2004) studied the association between student absenteeism and classroom CO_2 in Washington and Idaho. They found that 45% of classrooms studied "had short-term indoor CO_2 concentrations above 1000 ppm." They also found a statistically significant association between higher indoor–outdoor differences in CO_2 and student absenteeism. Haverinen-Shaughnessy et al. (2011) measured CO_2 in 100 classrooms, inferred that 87 had substandard ventilation rates, and found a positive association between the inferred ventilation rates (0.9–7.1 L/s per person) and student performance on standardized tests. Seppänen et al. (1999) reviewed the literature on ventilation rates, CO_2 concentrations, and sick-building syndrome symptoms. They reported that "about half of the carbon dioxide studies suggest that the risk of sick building syndrome symptoms continues to decrease with decreasing [indoor] carbon dioxide concentrations below 800 ppm." Evaluating data from a study conducted by EPA of 100 US office buildings, Erdmann and Apte (2004) found "statistically significant, dose-dependent associations ($P < 0.05$) for combined mucous membrane, dry eyes, sore throat, nose/sinus congestion, sneeze, and wheeze symptoms" with the difference between indoor and outdoor CO_2 levels.

It is important to note that those associations do not demonstrate that CO_2 itself is harmful to public health at the levels ordinarily encountered indoors. It may be that the adverse effects reported result from some other contaminant whose concentrations correlate with those of indoor CO_2. Alternatively, because CO_2 is not only a product of metabolism but a biologic trigger to induce breathing, it is conceivable that levels of CO_2 encountered indoors have direct health consequences. Studies of health hazards of CO_2 exposure have tended to stress conditions in healthy young adults, such as submariners (Margel et al., 2003), astronauts (Manzey and Lorenz, 1998), and motorcycle riders (Bruhwiler et al., 2005). Studies are lacking of the potential health consequences of chronic or episodic exposures to increased CO_2 at levels below 5,000 ppm in the young, the elderly, and the infirm.

OUTDOOR SOURCES

Air pollutants of outdoor origin enter buildings with ventilation air. Depending on the pollutant and the building conditions, the indoor proportion of the outdoor pollutant level ranges from zero (perfect sheltering) to 100% (no benefit from being indoors). A building provides virtually no protection against CO_2 from the outdoors, for example, but does provide some protection against PM and O_3. Filters in the mechanical ventilation system of typical commercial buildings actively remove some portion of the particles in the air that passes through them (Hanley et al., 1994). Particles can also deposit onto indoor surfaces passively, and this phenomenon reduces the indoor proportion of outdoor particles (Riley et al., 2002). Ozone can be removed from ventilation air by using activated carbon (Bekö et al., 2009; Shair, 1981), but the use of activated carbon in building mechanical systems is not common. On the other hand, O_3 reacts rapidly with indoor surfaces and with selected chemicals in indoor air (most notably nitric oxide [NO] and terpenes). Consequently, the indoor level of O_3 attributable to its presence outdoors is reduced, commonly to 20–70% of the outdoor level (Weschler, 2000). However, O_3 reactions indoors also generate byproducts—including formaldehyde, acrolein, and ultrafine particles—that have potential adverse effects (Weschler, 2006). Those two air pollutants, PM and O_3, currently receive the most attention in outdoor air-pollution control policy, and urban environments are furthest from compliance with air-quality standards for them. The relationship of indoor to outdoor concentrations for PM and O_3 have complex characteristics but also have been fairly well studied.

Epidemiologic studies indicate that ambient concentrations of PM and O_3 are associated with substantial adverse health effects. It is thus important to consider what could happen to those pollutants in indoor air in a climate-change regime. In addition to PM and O_3, which are addressed in detail below, other ambient pollutants are worth discussing, at least briefly; several are summarized at the end of this section.

Particulate Matter of Outdoor Origin

Airborne particles are a complex pollutant class, with source attributes, atmospheric dynamics, and health consequences that vary with size and chemical composition. EPA has established and maintains health-based national ambient-air quality standards (NAAQSs) for PM in outdoor air. The standards are based on 24-h average and annual average mass concentrations of particles finer than 10 μm in diameter (PM_{10}) and finer than 2.5 μm in diameter ($PM_{2.5}$). Particles 2.5–10 μm in diameter are referred to as *coarse*, and particles smaller than 2.5 μm in diameter are termed *fine*. The NAAQSs for PM do not consider the chemical composition of the particles. Emerging evidence suggests that inhalation exposure to *ultrafine* particles, those smaller than 0.1 μm in diameter, also poses health risks in a manner that would not be well captured by the existing NAAQSs (Sioutas et al., 2005). In addition to the ambient overall PM mass-concentration standards, some chemical components (such as lead) that would be found primarily in the particle phase are regulated separately, either under the NAAQS or as hazardous air pollutants (EPA, 2010c).

PM in outdoor air is strongly associated with adverse health outcomes. After a comprehensive review, Pope and Dockery (2006) concluded that "the literature provides compelling evidence that continued reductions in exposure to combustion-related fine particulate air pollution as indicated by $PM_{2.5}$ will result in improvements in cardiopulmonary health." In a 2009 study, Pope et al. concluded that "a decrease of 10 μg m^{-3} in the concentration of fine particulate matter was associated with an estimated increase in mean (±SE) life expectancy of 0.61 ± 0.20 year ($P = 0.004$)."

Coarse particles (Brunekreef and Forsberg, 2005) and ultrafine particles (Oberdörster, 2001) have also been associated with adverse health effects that can be different from those of the fine-particle fraction.

Atmospheric particles may be classified as primary or secondary. Primary PM is emitted directly in the particle phase from sources. Secondary particles are formed in the atmosphere from the conversion of gaseous precursors to condensed-phase species. Coarse particles are mainly of primary origin and tend to be mechanically generated, for example, by abrasion. Soil dust, sea salt, and fragments of tires, roadways, and vehicle brakes are primary particles that are found mainly in the coarse mode. Coarse particles are commonly removed fairly rapidly (in minutes to hours) from the atmosphere by a combination of gravitational settling and inertial impaction on the earth's surface. Because of their short atmospheric lifetime, concentrations of coarse particles can be spatially heterogeneous, with elevated concentrations found near emission sources.

The fine mode is a mixture of primary and secondary particles. Much of the primary material results from combustion processes and consists of noncombustible impurities, such as trace metals in coal, or products of incomplete combustion, such as soot (which is largely elemental carbon). Most of the atmospheric fine-particle mass is associated with sizes greater than 0.1 μm. Important secondary contributions to fine particles are associated with emissions of gaseous ammonia (NH_3), nitrogen oxides (mainly NO), and SO_2. Atmospheric oxidation processes convert the nitrogen oxides (NO_x) and SO_2 to nitric acid and sulfuric acid, which can then combine with NH_3 to form salts, such as ammonium nitrate or ammonium sulfate. These salts condense onto pre-existing particles to contribute to the fine-particle mass concentration.

Another important source of secondary PM derives from the emission of VOCs and SVOCs. Atmospheric oxidation processes tend to increase the polarity and reduce the vapor pressure of those species, causing their partitioning to shift from the gas phase to the condensed phase. Secondary particle formation occurs on a regional scale, in part because the relatively long time (hours to days) required for the atmospheric transformation processes allows substantial transport and dispersal from local and urban sources.

The particle size range 0.1–2 μm in diameter is also known as the *accumulation mode* because of the relatively long atmospheric persistence associated with these particles (Nazaroff, 2004). They are too big to diffuse and too small to settle rapidly, so they persist for many days in the atmosphere.[4] The combined importance of secondary formation and the slow

[4] Incorporation into cloud drops that precipitate is a major atmospheric removal mechanism that provides a typical atmospheric lifetime of 1–2 weeks for accumulation-mode particles.

atmospheric removal processes mean that fine-particle mass concentrations exhibit a higher degree of spatial homogeneity than do coarse particles.

Atmospheric ultrafine particles have important primary sources, mainly tailpipe emissions from internal-combustion engines (Kittelson, 1998). They are also formed through secondary nucleation events in the atmosphere (Kulmala et al., 2004). Ultrafine particles have relatively short atmospheric lifetimes. Primary ultrafine particles exhibit high spatial heterogeneity with very high concentrations on and near heavily traveled roadways. In contrast, secondary ultrafine particles are formed on a regional scale and so exhibit more spatial homogeneity. However, the secondary formation events occur as bursts, so the temporal variability associated with secondary ultrafine particles can be high.

The degree to which particles of outdoor origin are present indoors depends on three main factors: particle size, building ventilation rate, and the presence and degree of effectiveness of any filters used for removing particles from an HVAC system or from recirculated air. Particles in the accumulation mode have the greatest ability to penetrate and persist indoors (Bennett and Koutrakis, 2006; Nazaroff, 2004; Riley et al., 2002). Coarse particles have a more difficult time in penetrating infiltration cracks in the building envelope (Liu and Nazaroff, 2001) or penetrating fibrous filters in ventilation systems (Hanley et al., 1994). Coarse particles also deposit more rapidly onto indoor surfaces (Thatcher et al., 2002). Similarly, ultrafine particles penetrate infiltration cracks less effectively, are filtered more easily, and deposit onto indoor surfaces more rapidly (Lai and Nazaroff, 2000) than do accumulation-mode particles. Using a material balance model with empirical data on governing factors, Riley et al. (2002) estimated that an urban residence with a typical ventilation configuration would have indoor proportions of outdoor particles of about 0.45 for particle-number concentration (mainly ultrafine particles), about 0.8 for $PM_{2.5}$ mass concentration, but only about 0.2 for coarse-particle mass concentration. Those results also illustrate that buildings typically provide occupants some protection, but not extensive, from exposure to particles in outdoor air.

Absent active filtration, higher ventilation rates tend to produce higher indoor concentrations of outdoor particles. The reason is that ventilation serves as the sole source introducing outdoor particles into indoor air but as only one of several removal mechanisms. Higher ventilation rates increase the source term proportionally, but removal rates less than proportionally. Furthermore, higher rates of ventilation provided by open doors and windows (natural ventilation) tend to allow penetration with little attenuation.

In mechanically ventilated buildings, filters are commonly used to treat the supply air. Their efficiency can vary widely, as classified by the "minimum efficiency reporting value" (ASHRAE, 1999). The effectiveness of filtration in providing protection indoors against particles of outdoor origin depends both on filtration efficiency and on the airflow configuration.

In addition to those considerations, the chemical composition of particles can influence the penetration and persistence of outdoor particles in indoor environments. For example, Lunden et al. (2003) have shown that, under wintertime conditions, aerosol ammonium nitrate levels can be much lower indoors than outdoors. With warmer temperatures indoors, ammonium nitrate has an enhanced tendency to dissociate to its constituent gases, NH_3 and HNO_3, and the HNO_3 is then rapidly scavenged by the chemically basic gypsum wallboard commonly found indoors.

Regarding vulnerable populations, the findings of Hystad et al. (2009) should be noted. They found that "residences with low [economic] building values had higher infiltration efficiencies than other residences, which could lead to greater exposure gradients between low and high socioeconomic status individuals than previously identified using only ambient $PM_{2.5}$ concentrations." Results from McCormack et al. (2008) are also noteworthy in this regard. They reported time-integrated measurements of particle levels (PM_{10} and $PM_{2.5}$) for three-day periods in the homes of 300 children (ages 2–6) in Baltimore's inner city. The children were primarily African-American and from lower socioeconomic conditions. Smoking prevalence in the homes was 56%. Average indoor $PM_{2.5}$ and PM_{10} levels were higher indoors (39.5 and 56.2 μg/m^3, respectively) than the simultaneously measured outdoor levels (15.6 and 21.8 μg/m^3, respectively). Evidently because of the importance of indoor sources, open windows were associated with significantly lower indoor PM levels.

The question of what might be expected with respect to indoor particles of outdoor origin in a future in which climate change occurs is best addressed in two parts: What is expected to happen to outdoor particles? How might the indoor proportion of outdoor particles shift because of changes in building design and operation?

Regulated Particulate Matter

With respect to the first question, it is useful to consider the possibilities sorted into several categories of outdoor particles. Particles in outdoor air are subject to air-pollution control regulations. Given the strong regulatory, public-policy, and technology momentum and given that many areas in the United States are out of compliance with existing NAAQSs for PM, one might expect some overall improvement over the coming decades with regard to ambient particle levels, at least for the PM_{10} and $PM_{2.5}$ particle mass concentrations for which the regulatory machinery is the strongest.

- *Sulfate from coal-fired power plants.* Important contributions to improved ambient particle levels could be achieved by reducing SO_2 emissions from coal-fired power plants. A transformation in the direction of lower sulfur emissions from coal combustion could be accelerated because of climate-change concerns, in that coal-fired electricity in the United States accounts for a large proportion (5–10%) of global anthropogenic emissions of fossil carbon to the atmosphere.
- *Tailpipe emissions from motor vehicles.* A shift away from the use of petroleum as a transportation fuel would also have important benefits for reducing ambient particle concentrations. Because of the proximity of urban roadways to buildings, tailpipe emissions from vehicles have a higher effectiveness in causing indoor-air pollutant exposure per unit mass emitted than do central-station power plants, which emit their pollutants from tall stacks, often on the edge of or remote from populous regions. As with coal-fired electricity, an effective response to climate change in the transportation sector might yield cobenefits in reducing indoor exposure to PM. For example, a shift from vehicles powered by internal-combustion engines to plug-in hybrid vehicles, to electric vehicles, or to fuel-cell–powered vehicles could lead to a substantial net reduction in outdoor particle levels near buildings and consequent improvements in IAQ.
- *Distributed electricity generation.* A trend may emerge toward more distributed generation of electricity in the form of combustion close to the point of use (Pepermans et al., 2005). There are potential efficiency benefits if electricity generation is combined with use of waste heat (for example, to heat water in buildings). A potential disadvantage is that such technologies risk being more poorly controlled than central-station power plants, so emissions per unit of useful energy output may be higher. Furthermore, and more importantly, the efficiency in causing exposure may grow markedly higher in moving from central-station to distributed power generation because emissions and people will be in closer proximity in the case of distributed generation (Heath and Nazaroff, 2007; Heath et al., 2006).
- *Residential wood combustion.* Climate-change concerns might lead to increased use of wood combustion and the burning of other contemporary carbon fuels for home heating. Most US home heating is accomplished by burning fossil fuels either directly (e.g., natural gas or fuel oil at the home site) or indirectly (e.g., by use of electricity that is generated from burning coal or natural gas). Mitigation strategies to reduce greenhouse gas emissions will aim to reduce

societal reliance on fossil fuels. Among the potential strategies could be increased use of renewable fuels such as wood and other biomass sources as a substitute for fossil fuels. Residential wood smoke is an important contributor to ambient particle levels in the winter in many communities (McDonald et al., 2000; Naeher et al., 2007). It is also possible that wood-combustion technologies could be improved to the point where excessive emissions are limited or avoided (Olsson and Kjallstrand, 2006; Ward et al., 2010). And, if improved wood-combustion technologies are not sufficiently effective, it may be that community concerns about the adverse health risks from wood smoke exposure would constrain any increase in use.

- *Wildfires.* Climate change is expected to increase the frequency of wildfires. Higher ambient temperatures combined with episodes of drought could lead to periods with a higher tendency for forests to burn. Park et al. (2007) have evaluated the importance of burning of biomass of all types as a source of fine PM in the US ambient atmosphere. They have estimated that wildfires, other fires, and residential and industrial biofuel use currently account for 20% (eastern United States) to 30% (western United States) of total observed fine-particle concentrations in outdoor air. Furthermore, they have estimated that annual carbon emissions from open fires were about twice as high (0.7–0.9 Tg per year) as those from biofuel use (0.4 Tg per year). They concluded that biomass burning is "an important contributor to US air quality degradation, which is likely to grow in the future." Spracklen et al. (2009) have estimated that the annual mean area burned in the western United States will be about 50% larger in 2050 than in 2000, owing to climate-change effects; they also predict increases in summertime organic carbon and elemental carbon aerosol concentrations over the western United States of 40% and 20%, respectively, with most of the change attributable to increased wildfire emissions. Because wood-smoke particles are primarily in the fine mode, ordinary indoor environments, especially residences, do not provide much protection from them. However, Barn et al. (2008) have shown in an experimental study that using a recirculating, high-efficiency filter indoors can provide some protection against exposure to wood smoke associated with forest fires.
- *Windblown dust.* Another projected effect of climate change is increased frequency of drought in semi-arid regions. If water resources become further strained, changes in water allocations could increase the dryness of land surfaces, for example owing to reduced irrigation of crops and declining reservoir or lake levels. Conditions

such as these would have a tendency to increase the emissions of windblown dust into the atmosphere. Results of several studies illustrate the nature of the concern. Chan et al. (2008) reported that Asian dust storms were associated with an increased frequency of emergency-room visits for ischemic heart disease, cerebrovascular disease, and chronic obstructive pulmonary disease (COPD). Kuo and Shen (2010) showed that $PM_{2.5}$ and PM_{10} levels in an office building increased during a dust storm. Hefflin et al. (1994) reported very high PM_{10} levels (more than 1,000 μg m^{-3}) during seasonal dust storms in southeastern Washington state. However, on the basis of daily emergency-room visits, they concluded that, "the naturally occurring PM_{10} in this setting has a small effect on the respiratory health of the population in general." In contrast, Ostro et al. (2000) studied daily mortality in relation to particulate air pollution in the Coachella Valley, California, where "coarse particles of geologic origin are highly correlated with and comprise approximately 60% of PM_{10}, increasing to >90% during wind events." Their results demonstrated "associations between several measures of particulate matter and daily mortality in an environment in which particulate concentrations are dominated by the coarse fraction." Malig and Ostro (2009) assessed mortality statistics in 15 California counties for 1999–2005 in relation to coarse-particle monitoring data and found "evidence of an association between acute exposure to coarse particles and mortality" and that "lower socioeconomic status groups may be more susceptible to its effects."

Indoor Proportion of Outdoor Particles

The building stock in the future may substantially differ from current conditions. The body of evidence is weak for predicting how such changes may affect the infiltration and persistence of particulate matter from outdoor air. The basis is even weaker for attributing a portion of whatever evolution occurs specifically to climate change. Available information on conditions in the United States indicates that residential buildings have tended to become more airtight (Chan et al., 2005), which reduces air infiltration rates. Measurements in new single-family dwellings in California suggest that low ventilation rates are common in that portion of the building stock: 67% of 108 homes monitored had ventilation rates lower than the California building-code requirement of 0.35 air change per hour (Offermann, 2009). As noted before, lower air-exchange rates tend to provide some protection for building occupants against particles of outdoor origin. However, with lower air-exchange rates, concentrations of particles and other pollutants emitted from indoor sources would be higher.

The US housing stock seems to be moving toward more widespread use of mechanical systems to provide ventilation (Offermann, 2009; Russell et al., 2007). Mechanical systems that provide supply air can be equipped with filters to remove particles, and high filtration efficiency is available at modest cost (Bekö et al., 2008; Fisk et al., 2002). On the other hand, exhaust-only systems, such as continuous bathroom exhaust fans, do not provide the opportunity to deliberately filter supply air as a means of protecting occupants from outdoor particles. Furthermore, there are concerns that the presence of used filters in ventilation supply systems contributes to degraded IAQ and, for example, may increase the occurrence of sick-building syndrome symptoms (Bekö, 2009). Further technological innovation might be warranted to achieve economical and reliable high-performance mechanical ventilation systems in residences that provide good protection for occupants against particles of outdoor origin.

Ozone and Its Byproducts

Ozone is a secondary pollutant that is formed in the atmosphere by photochemical reactions involving NO_x and VOCs. Ozone concentrations in outdoor air have declined slowly in the United States, resisting relatively vigorous efforts to control precursor emissions. The background level of O_3 in the clean troposphere also has risen. As health-science information has improved, the NAAQS standard for O_3 has become more stringent.

Several modeling studies have explored the consequences of climate change for outdoor O_3 concentrations. Hogrefe et al. (2004b) combined a global circulation model, a mesoscale regional climate model, and an air quality model to simulate summertime ozone levels in the eastern United States for the 2020s, 2050s, and 2080s. As compared with the 1990s, and considering only the effects of climate change (using the Intergovernmental Panel on Climate Change's A2 scenario), the maximum 8-h ozone level increased by 2.7, 4.2, and 5.0 ppb, respectively, for the three future time periods. In a related study, Bell et al. (2007) estimated hourly concentrations in 50 eastern US cities in the 1990s and also for the 2050s, taking account of the predicted change in climatic conditions (again using International Panel on Climate Change [IPCC] Scenario A2) but not accounting for changes in anthropogenic precursor emissions. A key finding of their study is that "on average across the 50 cities, the summertime daily 1-h maximum [O_3 level] increased 4.8 ppb, with the largest increase at 9.6 ppb." Tagaris et al. (2009) reported on the results of a detailed modeling study of outdoor $PM_{2.5}$ and O_3 levels in the United States. Like the study by Bell et al., this study did not account for changes in emission sources or population. Tagaris et al. estimated that climate-change–induced shifts in $PM_{2.5}$ levels would cause roughly 4,000 additional deaths per year in 2050 compared with 2001, and 300 additional deaths per year would be caused by increasing O_3 concentrations. In an earlier study, Tagaris et al. (2007) reported model predictions of regional concentrations of O_3 and $PM_{2.5}$ over the whole United States, incorporating not only the direct effects of climate change (using IPCC Scenario A1B) but expected emission reductions for the year 2050. They estimated that emitted NO_x and SO_2 would be reduced by more than 50%. They found that "impacts of global climate change alone on regional air quality are small compared to impacts from emission control-related reductions." Overall, they predicted a 20% decrease in the mean summer maximum daily 8-h O_3 levels and that mean annual $PM_{2.5}$ levels would be an average of 23% lower. Racheria and Adams (2009) published an analogous study in which they concluded that "climate change, by itself, significantly worsens the severity and frequency of high O_3 events over most locations in the US, with relatively small changes in average O_3 air quality."

Buildings offer some protection from O_3 exposure because O_3 irreversibly decomposes on indoor surfaces and reacts with some gas-phase species (primarily NO and terpenes) that may be found indoors. However, some O_3 that penetrates does persist. Given common residual O_3 levels indoors and the fact that people spend most of their time indoors, most O_3 exposure occurs indoors (Weschler, 2006). New evidence from research on O_3-initiated chemistry raises a potentially important question: To what extent are the health risks that are ascribed to ozone exposure influenced by the coincident exposure of the products of ozone-initiated chemistry? Ozone-initiated chemistry, producing potentially health-relevant volatile byproducts such as aldehydes and organic acids, can occur on indoor surfaces (Weschler, 2004), on clothing (Coleman et al., 2008), on hair (Pandrangi and Morrison, 2008), and even on human skin (Wisthaler and Weschler, 2010).

The distinction is important in the context of climate-change effects on IAQ and health. Changes in building design and operation can be anticipated owing to development of new materials, resource limitations, changing economic conditions, changing fashion, and other factors (Weschler, 2009). Such changes might deliberately or inadvertently alter the indoor–outdoor relationship for O_3, for example, through the introduction of active or passive controls (Kunkel et al., 2010; Lee and Davidson, 1999; Shair, 1981). Such changes could also deliberately or inadvertently alter the nature, degree, and importance of O_3-initiated indoor chemistry. These considerations overlap but are not coincident. Overall, if ambient O_3 concentrations increase while ventilation rates decrease, the net effect on indoor O_3 concentrations is uncertain, because changes in these two factors have opposing influence on indoor ozone levels. However, both of these factors tend to increase the indoor concentrations of the byproducts of O_3-initiated chemistry.

Pollen

Researchers have suggested that pollen levels in outdoor air may rise as a consequence of higher CO_2 levels, warmer temperatures, and concomitant longer growing seasons resulting from climate change (Ziska et al., 2009), which would have consequences for health outcomes like allergic rhinitis, asthma, and atopic dermatitis (Reid and Gamble, 2009). Intact pollen grains are relatively large (a few tens of micrometers in diameter). Thus, they should neither effectively penetrate into nor persist in indoor air (Liu and Nazaroff, 2001; Nazaroff, 2004; Sippola and Nazaroff, 2003). Nor should they penetrate further into the respiratory system than the head if inhaled (Yeh et al., 1996). Consideration of those factors suggests that buildings would provide good protection against whole pollen grains and that the biologic insult associated with exposure to whole grains should be concentrated in the extrathoracic regions (eyes, nose, and throat). The tracking of pollen grains into buildings (for example, on clothing) might constitute an IAQ and health concern if the grains are later resuspended indoors. Furthermore (and perhaps more important), pollen grains can fracture, generating much smaller particles (0.5–3 μm in diameter) (D'Amato et al., 2007) that carry allergenic proteins. The smaller particles could penetrate both the building envelope and the upper respiratory tract.

Ziska et al. (2011) studied the effects of the rise of frost-free days during 1995–2009 on the length of the ragweed (*Ambrosia* spp.) pollen season in the United States. They found that the duration of the season increased by 13–27 days at latitudes greater than about 44°N.[5] They noted that longer pollen seasons and higher exposure to pollen may intensify allergic sensitization and increase the duration and severity of allergy symptoms. The committee did not identify any literature specifically regarding climate change and indoor exposure to pollen.

Algal Blooms After Floods

Harmful algal blooms (HABs) occur when saltwater or freshwater reservoirs accumulate algae or other protozoa to abundances at which their biomass or toxins lead to adverse effects on aquatic life or humans. Human activity can affect the frequency and severity of HABs, for example, through

5 The 44th parallel passes through the northern United States, including Oregon, Idaho, Wyoming, South Dakota, Minnesota, Wisconsin, Michigan, New York, Vermont, New Hampshire, and Maine.

increased fertilizer runoff and aquaculture that leads to eutrophication in rivers and coastal areas. Climate also affects the appearance and distribution of HABs. Two factors with known effects are change in water temperature and changes in nutrient levels. Changing patterns of the types of species involved in HABs and their timing can occur with increases in mean water temperature (Glibert et al., 2005). Increased Saharan dust storms have been shown to be rich in iron, a limiting nutrient, which can lead to increases in some species of algae that proliferate in the Caribbean (Lenes et al., 2001).

The red-tide alga, *Kernia brevis*, produces brevetoxin. These cyclic polyether molecules "become part of the marine aerosol as the fragile, unarmored cells are broken up by wave action. Inhalation of the aerosolized toxin results in upper and lower airway irritation" (Milian et al., 2007). In Florida, increased respiratory irritation has been reported almost annually during red-tide events. Researchers have found that residents, lifeguards, and tourists report many more respiratory symptoms after exposure to red-tide events (Backer et al., 2003, 2005; Fleming et al., 2005). During two red-tide exposure periods in 2005 and 2006, a cohort of asthmatic children and adults in Sarasota, Florida, was studied (Fleming et al., 2009). Their exposure to brevetoxin was assessed via personal air sampling, and their symptoms via questionnaire (Cheng et al., 2010). Researchers observed associations between brevetoxin exposure and increased respiratory symptoms. Other researchers have also reported increases in emergency-department visits for asthma, pneumonia, and bronchitis in residents during red-tide events (Kirkpatrick et al., 2006). There is also evidence that brevetoxin can affect the mucociliary escalator in animals (Abraham et al., 2005). Therefore, exposure to brevetoxin in red-tide events could affect the respiratory tract's ability to clear other inhaled particles, such as allergens, endotoxins, and fungal spores. The ramifications would be increased residence time of the particles in the airways and a higher biologically relevant dose. A monitoring system is in place for red-tide events and other HABs, but it is mainly for ensuring seafood safety; it is not linked to respiratory health protection.

Flooding caused by extreme precipitation events, which may increase in number and severity under climate change conditions, are commonly followed by disease clusters (IOM, 2008) and may lead to both more frequent HABs and increased exposure to potentially harmful agents associated with them. The committee could not, however, identify any literature specifically addressing changes in risk associated with exposure to these agents in indoor environments.

Sulfur Dioxide

Ambient SO_2 is primarily a result of coal combustion and originates from the presence of sulfur as an impurity in coal. EPA data show that about 68% of nationwide atmospheric sulfur emissions in 2002 were from "electricity generating units" and that the other important sources were "industrial/commercial/residential fuels" (about 16%) and "industrial processes" (about 8%) (EPA, 2009a). Ambient NAAQSs for SO_2 and acid-rain legislation (in the 1990 Clean Air Act Amendments) have led to substantial reductions (about 50%) in SO_2 emissions from power plants. The largest remaining emissions are from older power plants whose high emission rates continue to be allowed. New coal-fired power plants are required to have good emission controls for SO_2 that are achieved, for example, with flue-gas desulfurization.

Future ambient SO_2 levels might rise or fall depending on changes in the use of coal as an energy source and on emission controls. The indoor environment provides some protection against SO_2 because, as an acid gas, it reacts on indoor surface materials (Biersteker et al., 1965; Grøntoft and Raychaudhuri, 2004; Walsh et al., 1977).

Nitrogen Oxides

NO_x (mainly NO and NO_2) are emitted primarily as a result of combustion. To some extent, the presence of nitrogen in fuel (as in coal) leads to NO_x emissions. However, any high-temperature combustion process that uses air as the oxidizer can also produce NO_x emissions, with the nitrogen originating from N_2 in the combustion air. Important sources of NO_x in ambient air are mobile sources (both on-road and off-road), fossil-fueled power plants (using coal and natural gas), and other stationary combustion of (mainly) fossil fuels. For 2002, EPA national emission inventory data indicate that mobile sources were responsible for about 60% of NO_x emissions. Because NO_x is a precursor of O_3 and other photochemical smog components, it has been and continues to be subjected to strong emission-control efforts, and continuing progress in reducing emissions in the near future can be expected. A high level of scrutiny and emission control is expected especially for diesel emissions, which are becoming progressively more important (Dallmann and Harley, 2010). Less future reliance on fossil fuels in particular and combustion in general suggests that NO_x emissions may also decrease in a future climate-change regime. The indoor environment provides modest to moderate protection against NO_2 of outdoor origin (Quackenboss et al., 1986).

Hazardous Air Pollutants

There is a long list of species known as HAPs. In the United States, about 190 HAPs were designated under the 1990 Clean Air Act Amendments. In contrast to the *criteria pollutants*,[6] HAPs are regulated only with respect to emissions from major sources; there are no ambient concentration standards, and the concentrations of these pollutants are not routinely monitored. However, summary appraisals have combined emissions data with dispersion modeling and risk factors to discern which pollutants are most prevalent and where the health risks posed by HAPs are highest. For example, in one study, the median hazard ratio (average ambient concentration divided by a cancer benchmark value) was highest for 1,3-butadiene, formaldehyde, benzene, carbon tetrachloride, chromium, methyl chloride, and chloroform (Woodruff et al., 1998). For chronic noncancer toxicity, acrolein had the highest median hazard ratio in the study. Studies investigating the indoor–outdoor relationships of HAPs reveal that for many species in many buildings indoor concentrations are higher than those outdoors (e.g., Jia et al., 2008a). That characteristic demonstrates the importance of indoor emission sources in contributing to indoor levels. On the other hand, levels of indoor HAPs that are attributable to their presence in outdoor air have not been well studied. For some important species, such as benzene and the chlorinated organics, it is reasonable to expect that indoor environments provide little or no protection from outdoor concentrations. For other species that have higher reactivity, such as acrolein and aldehydes, the penetration and persistence of outdoor pollutants into indoor environments is not known. Future trends in the outdoor levels of the pollutants in a climate-change regime are not clear, but the scrutiny that they are receiving as HAPs suggests that emissions might decline.

INDOOR AIR QUALITY IN DEVELOPING COUNTRIES

One of the dominant environmental health concerns in developing countries results from the use of solid biomass fuels for cooking. Around 2.7 billion people are thought to rely on burning of biomass (dung, crop residues, and wood) or coal as their household fuel, and about 82% of them live in rural areas (IEA/UNDP/UNIDO, 2010). In sub-Saharan Africa, around 80% of the population relies on the traditional use of biomass for cooking, and the electrification rate is only 31%. Combustion of such fuels in open fires or cookstoves is generally inefficient and leads to very high concentrations of products of incomplete combustion, which have

[6] The criteria pollutants defined under the NAAQS are O_3, particulate matter, CO, NOx, SO_2, and lead.

serious health implications for those exposed. The products of incomplete combustion form a complex mixture of pollutants, including fine PM, NO_2, sulfur oxides (particularly in the case of coal), CO, and polycyclic aromatic compounds, such as benzo[a]pyrene (Smith, 1993). Many of the compounds are known to pose health hazards. Research also has demonstrated that indoor air pollution from solid biomass fuels may contribute to climate change as a result of emissions of black carbon, methane, CO, and nonmethane VOCs (which are O_3 precursors). The overall effect depends on the balance between warming aerosols, such as black carbon, and other types of particles that may be cooling; the net effect is a likely contribution to warming. Black carbon may also accelerate the melting of glaciers when it is deposited in mountainous areas, such as the Himalayas (Ramanathan and Carmichael, 2008). Even if the fuel contains contemporary carbon (for example, the carbon associated with crop residue, wood, or animal dung), there may be a greater climate-change effect than would be the case for an efficient stove burning a fuel that contains fossil carbon, such as liquefied petroleum gas. Improving the cooking conditions for this large population offers the potential for cobenefits: improved public health and reduction in climate-change effects. Wilkinson et al. (2009) quantify the cobenefits for rural India, demonstrating large potential benefits in improved public health and not insignificant improvements in climate-change effects.

Exposures to such pollutants as PM in houses that burn biomass or coal tend to be very high. For example, PM levels can be one or even two orders of magnitude above the EPA NAAQSs, depending on type of fuel, stove characteristics, and housing.

Exposure to air pollution associated with indoor use of solid fuels has been implicated, with various degrees of certainty, as a causative factor in several adverse health outcomes. In total, it is thought that around 1.6 million deaths a year can be attributed to indoor air pollution of this type globally (Ezzati et al., 2002).

Increased incidences of acute respiratory infections in children and of chronic obstructive airways disease, particularly in women, comprise the most compelling evidence for adverse health effects (Bruce et al., 2000). Exposure to these types of indoor air pollution, particularly when coal is used, may also cause lung cancer in women. In low- to middle-income countries—such as China, India, and Mexico—two-thirds of women who have lung cancer are nonsmokers. Exposure to so-called smoky coal seems to be strongly related to lung cancer in China (Mumford et al., 1995). The situation is less clear with regard to wood-smoke exposure, although exposure to known carcinogens, such as benzo[a]pyrene, is likely to be equivalent to smoking several cigarettes per day, so such a health risk cannot be ruled out. There is also some epidemiologic evidence of a causal association between indoor exposure to smoke and upper aerodigestive tract[7] cancers in Latin America (Pintos et al., 1998).

There is reasonably consistent evidence of an association between particulate air pollution and both hospital admissions and cardiovascular-disease mortality, but the association specifically with myocardial infarction is less consistent (Bhaskaran et al., 2009). And although it is plausible that indoor air pollution increases the risk of cardiovascular disease events and mortality, there is a lack of direct evidence (Smith, 2000).

Low birth weight has been associated with use of wood fuel by mothers in Guatemala, perhaps mediated through CO exposure (Boy et al., 2002), and there may be an association with perinatal mortality. Several studies in India have shown an association between biomass fuel and cataracts (cortical, nuclear, and mixed but not posterior subcapsular in one study) (Mohan et al., 1989), and animal studies support this association. A possible association between exposure to wood smoke and tuberculosis has been indicated in a few studies (e.g., Mishra et al., 1999), but a 2010 review concluded that "available original studies looking at this issue do not provide sufficient evidence of an excess risk of tuberculosis due to exposure to indoor coal or biomass combustion" (Slama et al., 2010).

Wilkinson and colleagues (2009) report that advanced biomass stoves available in India at prices of US$20–50 can achieve around 15 lower particle emissions per meal compared with traditional stoves. Hybrid gasifier stoves with small electric blowers can achieve good performance with a range of fuel characteristics but require a source of electricity. There has been experience with large national stove programs, such as the provision of 180 million improved stoves in China over 12 years starting in 1983. A modeling exercise indicated that a 150 million–stove program in India, to be implemented over 10 years, that would provide improved-efficiency cookstoves through various delivery mechanisms, including to poor women receiving antenatal care, could result in the prevention of around 2.2 million premature deaths from acute respiratory infections in children and from COPD and ischemic heart disease in adults (Wilkinson et al., 2009). However, because of the slow evolution of COPD, there may be a substantial lag between the introduction of the new stoves and the reduction in deaths. Wilkinson and colleagues also estimated that such a program could result in a reduction in greenhouse pollutants equivalent to 0.5–1.0 billion tons of CO_2 over the decade. Many other benefits could accrue, such as reductions in time spent looking for fuel, in expenditure for fuel, and, in some locations where wood is not harvested sustainably, in deforestation.

[7] The aerodigestive tract comprises the upper respiratory and digestive tracts, including the tissues associated with the nose, lips, tongue, mouth, throat, vocal cords, and upper portions of the esophagus and windpipe.

A randomized trial in Guatemala of improved woodstoves with chimneys compared with traditional open fires showed encouraging preliminary results, particularly with regard to acute respiratory infections in infants up to 18 months old, but did not have the power to assess differences in mortality or birth weight (WHO, 2007). Other larger trials were planned or under way at the time this study was completed.

Although levels of indoor air pollutants may be lowered by the use of biogas or liquefied petroleum gas (LPG) or, of course, by provision of electricity, these are all more expensive alternatives and may not be suitable for some communities. The development and implementation of culturally acceptable highly efficient biomass cookstoves will be needed for the indefinite future to meet the requirements of many poor communities for inexpensive, clean, and convenient sources of household energy. Such efforts could yield cobenefits in contributing to the mitigation of climate change.

There is little published information about how climate change might affect IAQ in low-income countries. In theory, climate change could affect the availability of biomass fuel, but the direction and magnitude of the change may depend on the region and on the relative effects of carbon dioxide fertilization and changes in temperature and precipitation. More work is also needed on how changes in forest management, such as the trend toward greater use of forest plantations to supply roundwood, and the commercial exploitation of crops for biofuel, may affect the availability of biomass for use by poor households and thus indirectly affect the indoor environment.

CONCLUSIONS

On the basis of its review of papers, reports, and other information presented in this chapter, the committee has reached the following conclusions regarding the effects of climate change on IAQ and its consequent influence on public health.

- The elements that influence important outcomes in the climate-change–IAQ–public-health nexus are numerous and diverse. They are interconnected in a complex manner that includes feedback loops and interweaves natural processes with technology, individual human behavior, and social systems. It is those systemic features, rather than the nature of individual elements themselves, that pose the greatest challenges for understanding and effectively addressing the effects of climate change on IAQ and public health.
- We understand relatively little about how climate change will affect IAQ and thereby public health. More information is available on the factors that influence indoor concentrations of health-relevant

pollutants and how concentrations might shift as a consequence of climate change.

- Three classes of factors have important influences on indoor pollutant concentrations: pollutant attributes (including source properties), building characteristics, and human behavior.
- The concentration of any indoor air pollutant in any indoor space is governed by a balance between emissions and removal. Concentrations in combination with human occupancy govern exposures. Excessive exposures confer health risks on those exposed. Climate change can affect this system in numerous particular ways. For example, by causing an increase in the outdoor concentrations of some pollutants in some places and at some times, climate change is likely to increase indoor concentrations and associated exposures in buildings in those places and at those times.
- In addition to direct shifts caused by climate change, there are likely to be shifts in IAQ that are mediated by human responses to climate change. For example, mitigation measures to reduce energy use in buildings could lead to systematically lower ventilation rates in buildings that would cause higher concentrations of and exposures to pollutants that are emitted from indoor sources. Another example is the increased use of air conditioning, an expected adaptation measure, which could exacerbate anthropogenic emissions of greenhouse gases and, if accompanied by reduced ventilation rates, increase the indoor concentrations of pollutants emitted from indoor sources.
- Reactions to weather emergencies pose public-health risks related to IAQ, such as the potential for poisoning from exposure to CO emitted from emergency electricity generators. Such emergencies may increase in frequency if climate change results in more frequent or more severe climate events like storms.
- Actions taken by individuals can profoundly influence the IAQ in individual buildings. There is a public interest as well as an individual interest in seeing that the system for establishing and maintaining good IAQ works well. Negligent or ill-informed behavior by individuals can cause serious harm.
- Indoor pollutant concentrations can be separated into contributions from indoor sources and from outdoor air. Combustion is a major source of both outdoor and indoor air pollution and is arguably the most important source of indoor air pollutants with respect to health risks. Important combustion-related issues associated with indoor emissions are CO exposures from emergency-generator use and IAQ problems associated with cooking, heating, smoking, and small-scale activities, such as use of candles and in-

cense. Other important pollutants associated primarily with indoor sources include radon, VOCs, and SVOCs. Outdoors, the main pollutants of concern are PM and O_3. Specific PM concerns that may be exacerbated by climate change include increases in smoke from wild-land fires, pollen, and windblown dust.

- There is a large gap between what is known about IAQ and public health and what needs to be known. However, the gap between what is known and what is done to address problems is even larger. One of the risks associated with how climate change will affect IAQ is that the gap between what is known and what is done will grow and have adverse consequences for public health.
- In formulating responses to the challenges posed to IAQ and public health by climate change, it is important to recognize and account for the diversity in subpopulations, in part because of variability in susceptibility to the effects of indoor air-pollutant exposure. It is also important to take account of variability within populations in the knowledge and resources with which to take effective action in response to changing conditions.
- It is important to take proper account of the different issues of concern and appropriate responses for different building types, such as single-family dwellings, multifamily apartment buildings, schools, health-care facilities, and offices.
- Efforts to save energy by improving building performance need to be accompanied by strong caution with respect to changing building ventilation rates. Two driving forces are apparent. First, as a mitigation measure, efforts to save energy in buildings are gaining momentum. Energy is required to condition the temperature and humidity of ventilation air, so individuals and organizations may seek to save energy by reducing the rate of ventilation of indoor spaces. Second, as temperatures rise during the warm parts of the year, there may be a progressive shift to greater reliance on air conditioning and away from cooling by means of open windows. The effect of ventilation on IAQ has multiple facets that operate in different directions, so one cannot be certain a priori of the net effect for each building. A lower building ventilation rate will tend to provide enhanced protection against some pollutants from outdoors, such as PM. But reduced ventilation rates tend to cause concentrations of pollutants that originate primarily from indoor sources to increase. Reducing ventilation rates does not automatically mean that IAQ problems will become worse; nor is it appropriate to assume that no problems will be associated with reducing ventilation rates.

- Attention to controlling indoor emission sources is warranted. For many pollutants with indoor sources, it has been found that variability in emissions, not variability in ventilation rates, is the primary determinant of whether indoor air-pollutant levels are excessive. Put another way, when the indoor emission rates are high, ventilation in a normal range is unlikely to be sufficient to avoid a problem. There is no evidence that clearly links increased indoor pollutant emission rates to climate change. However, there are several potential concerns that deserve attention, including CO from emergency generators, emissions from cooking, emissions from heating systems (including unvented combustion appliances, backdrafting, and increased use of wood as a fuel), emissions from smoking, emissions from use of candles, radon from intimate contact of indoor spaces with earthen materials, and VOCs and SVOCs from various indoor sources. Special attention is needed to ensure that life-cycle impact assessments aimed at improving the environmental performance of buildings take proper account of the disproportionately large effects that emissions from indoor materials and processes can have on IAQ and public health.
- Attention ought to also be directed toward improving understanding of the effectiveness of indoor environments as a shelter against pollutants of outdoor origin that may be altered owing to climate change. To date, the scientific literature on the affects of climate change on outdoor air quality has focused on criteria pollutants, especially $PM_{2.5}$ and O_3. There are good regulatory and technologic systems in place that are striving to reduce emissions from anthropogenic sources. The momentum is expected to continue to yield improvements in reducing ambient pollutant concentrations that are clearly associated with anthropogenic sources. A greater concern would apply to pollutants that lie outside the regulatory structure of managed emissions, such as smoke from wildfires, pollen from weeds, and windblown dust. Pollutant emissions from sources like these might be substantially worsened by climate change. If so, indoor environments will be used as imperfect shelters that could be improved with proper attention and a commitment of appropriate resources.

REFERENCES

Abraham WM, Bourdelais AJ, Sabater JR, Ahmed A, Lee TA, Serebriakov I, Baden DG. 2005. Airway responses to aerosolized brevetoxins in an animal model of asthma. *American Journal of Respiratory and Critical Care Medicine* 171:26-34.

Abt E, Suh HH, Catalano P, Koutrakis P. 2000. Relative contribution of outdoor and indoor particle sources to indoor concentrations. *Environmental Science & Technology* 34:3579-3587.

Ad-hoc Work Group (Ad-hoc-Arbeitsgruppe). 2008. Gesundheitliche bewertung von kohlendioxid in der innenraumluft [Health evaluation of carbon dioxide in indoor air]. *Bundesgesundheitsblatt–Gesundheitsforschung–Gesundheitsschutz* 51:1358-1369.

Alcock RE, Sweetman A, Prevedouros K, Jones KC. 2003. Understanding levels and trends of BDE-47 in the UK and North America: An assessment of principal reservoirs and source inputs. *Environment International* 29:691-698.

Apple J, Vicente R, Yarberry A, Lohse N, Mills E, Jacobson A, Poppendieck D. 2010. Characterization of particulate matter size distributions and indoor concentrations from kerosene and diesel lamps. *Indoor Air* 20(5):399-411.

ASHRAE (American Society of Heating, Refrigerating and Air Conditioning Engineers). 1999. *Method of testing general ventilation air-cleaning devices for removal efficiency by particle size.* ASHRAE/ANSI Standard 52.2-1999. Atlanta, GA: American Society of Heating, Refrigerating and Air-Conditioning Engineers.

ASHRAE. 2010. *Ventilation for acceptable air quality.* ANSI/ASHRAE Standard 62.1-2010. Atlanta, GA: American Society of Heating, Refrigerating and Air-Conditioning Engineers.

ATSDR (Agency for Toxic Substances and Disease Registry). 2002. *Toxicological profile for di(2-ethylhexyl)phthalate.* Atlanta, GA: Agency for Toxic Substances and Disease Registry, US Department of Health and Human Services.

Backer LC, Fleming LE, Rowan A, Cheng YS, Benson J, Pierce RH, Zaias J, Bean J, Bossart GD, Johnson D, Quimbo R, Baden DG. 2003. Recreational exposure to aerosolized brevetoxins during Florida red tide events. *Harmful Algae* 2:19-28.

Backer LC, Kirkpatrick B, Fleming LE, Cheng YS, Pierce R, Bean JA, Clark R, Johnson D, Wanner A, Tamer R, Zhou Y, Baden DG. 2005. Occupational exposure to aerosolized brevetoxins during Florida red tide events: Effects on a healthy worker population. *Environmental Health Perspectives* 113:644-649.

Barn P, Larson T, Noullet M, Kennedy S, Copes R, Brauer M. 2008. Infiltration of forest fire and residential wood smoke: An evaluation of air cleaner effectiveness. *Journal of Exposure Science and Environmental Epidemiology* 18:503-511.

Bekö G, Clausen G, Weschler CJ. 2008. Is the use of particle air filtration justified? Costs and benefits of filtration with regard to health effects, building cleaning and occupant productivity. *Building and Environment* 43:1647-1657.

Bekö G. 2009. Used filters and indoor air quality. *ASHRAE Journal* 51(3):64-72.

Bekö G, Fadeyi MO, Clausen G, Weschler CJ. 2009. Sensory pollution from bag-type fiberglass ventilation filters: Conventional filter compared with filters containing various amounts of activated carbon. *Building and Environment* 44:2114-2120.

Bekö G, Lund T, Nors F, Toftum J, Clausen G. 2010. Ventilation rates in the bedrooms of 500 Danish children. *Building and Environment* 45:2289-2295.

Bell ML, McDermott A, Zeger SL, Samet JM, Dominici F. 2004. Ozone and short-term mortality in 95 US urban communities, 1987-2000. *Journal of the American Medical Association* 292:2372-2378.

Bell ML, Goldberg R, Hogrefe C, Kinney PL, Knowlton K, Lynn B, Rosenthal J, Rosenzweig C, Patz JA. 2007. Climate change, ambient ozone, and health in 50 US cities. *Climatic Change* 82:61-76.

Bennett DH, McKone TE, Evans JS, Nazaroff WW, Margni MD, Jolliet O, Smith KR. 2002. Defining intake fraction. *Environmental Science & Technology* 36:206A-211A.

Bennett DH, Koutrakis P. 2006. Determining the infiltration of outdoor particles in the indoor environment using a dynamic model. *Journal of Aerosol Science* 37:766-785.

Bhangar S, Mullen NA, Kreisberg NM, Hering SV, Nazaroff WW. 2011. Ultrafine particle concentrations and exposures in seven residences in northern California. *Indoor Air* 21(2):132-144.

Bhaskaran K, Hajat S, Haines A, Herrett E, Wilkinson P, Smeeth L. 2009. Effects of air pollution on the incidence of myocardial infarction. *Heart* 95:1746-1759.

Biersteker K, DeGraaf H, Nass CAG. 1965. Indoor air pollution in Rotterdam homes. *Air and Water Pollution* 9:343-350.

Bohac DL, Hewett MJ, Hammond SK, Grimsrud DT. 2011. Secondhand smoke transfer and reductions by air sealing and ventilation in multi-unit buildings: PFT and nicotine verification. *Indoor Air* 21(1):36-44.

Bornehag CG, Sundell J, Weschler CJ, Sigsgaard T, Lundgren B, Hasselgren M, Hägerhed-Engman L. 2004. The association between asthma and allergic symptoms in children and phthalates in house dust: A nested case–control study. *Environmental Health Perspectives* 112:1393-1397.

Boy E, Bruce N, Delgado H. 2002. Birth weight and exposure to kitchen wood smoke during pregnancy in rural Guatemala. *Environmental Health Perspectives* 110:109-114.

Breysse PN, Buckley TJ, Williams D, Beck CM, Jo SJ, Merriman B, Kanchanaraksa S, Swartz LJ, Callahan KA, Butz AM, Rand CS, Diette GB, Krishnan JA, Moseley AM, Curtin-Brosnan J, Durkin NB, Eggleston PA. 2005. Indoor exposures to air pollutants and allergens in the homes of asthmatic children in inner-city Baltimore. *Environmental Research* 98:167-176.

Broder I, Corey P, Cole P, Lipa M, Mintz S, Nethercott JR. 1988. Comparison of health of occupants and characteristics of houses among control homes and homes insulated with urea formaldehyde foam. II. Initial health and house variables and exposure-response relationships. *Environmental Research* 45:156-178.

Broder I, Corey P, Brasher P, Lipa M, Cole P. 1991. Formaldehyde exposure and health status in households. *Environmental Health Perspectives* 95:101-104.

Brown SK, Sim MR, Abramson MJ, Gray CN. 1994. Concentrations of volatile organic compounds in indoor air—A review. *Indoor Air* 4:123-134.

Bruce N, Perez-Padilla R, Albalak R. 2000. Indoor air pollution in developing countries: A major environmental and public health challenge. *Bulletin of the World Health Organization* 78:1078-1092.

Bruhwiler PA, Stampfli R, Huber R, Camenzind M. 2005. CO_2 and O_2 concentrations in integral motorcycle helmets. *Applied Ergonomics* 36:625-633.

Brunekreef B, Forsberg B. 2005. Epidemiological evidence of effects of coarse airborne particles on health. *European Respiratory Journal* 26:309-318.

Buonanno G, Morawska L, Stabile L. 2009. Particle emission factors during cooking activities. *Atmospheric Environment* 43:3235-3242.

Butte W, Heinzow B. 2002. Pollutants in house dust as indicators of indoor contamination. *Reviews of Environmental Contamination & Toxicology* 175:1-46.

Butte, W. 2004. Sources and impacts of pesticides in indoor environment. In *The Handbook of Environmental Chemistry*, edited by Hutzinger O. Berlin: Springer-Verlag.

Cal/EPA (California Environmental Protection Agency). 2011. *Composite wood products ATCM*. http://www.arb.ca.gov/toxics/compwood/compwood.htm (accessed February 16, 2011).

Cal/EPA ARB (Air Resources Board). 2004. *Indoor air quality guideline: Formaldehyde in the home*. http://www.arb.ca.gov/research/indoor/formaldGL08-04.pdf (accessed May 3, 2011).

Casset A, Marchand C, Purohit A, le Calve S, Uring-Lambert B, Donnay C, Meyer P, de Blay F. 2006. Inhaled formaldehyde exposure: Effect on bronchial response to mite allergen in sensitized asthma patients. *Allergy* 61:1344-1350.

CDC (Centers for Disease Control and Prevention). 1997. Use of unvented residential heating appliances—United States, 1988-1994. *MMWR (Morbidity and Mortality Weekly Report)* 46(51):1221-1224.

CDC. 2005. *Third national report on human exposure to environmental chemicals.* Atlanta, GA: Centers for Disease Control and Prevention, National Center for Environmental Health.

Chan CC, Chuang KJ, Chen WJ, Chang WT, Lee CT, Peng CM. 2008. Increasing cardiopulmonary emergency visits by long-range transported Asian dust storms in Taiwan. *Environmental Research* 106:393-400.

Chan WR, Nazaroff WW, Price PN, Sohn MD, Gadgil AJ. 2005. Analyzing a database of residential air leakage in the United States. *Atmospheric Environment* 39:3445-3455.

Cheng YS, Zhou Y, Naar J, Irvin CM, Su WC, Fleming LE, Kirkpatrick B, Pierce RH, Backer LC, Baden DG. 2010. Personal exposure to aerosolized red tide toxins (brevetoxins). *Journal of Occupational and Environmental Hygiene* 7:326-331.

Chevrier J, Eskenazi B, Holland N, Bradman A, Barr DB. 2008. Effects of exposure to polychlorinated biphenyls and organochlorine pesticides on thyroid function during pregnancy. *American Journal of Epidemiology* 168(3):298-310.

Chiang TA, Wu PF, Wang LF, Lee H, Lee CH, Ko YC. 1997. Mutagenicity and polycyclic aromatic hydrocarbon content of fumes from heated cooking oils produced in Taiwan. *Mutation Research—Fundamental and Molecular Mechanisms of Mutagenesis* 381:157-161.

CMHC (Canada Mortgage and Housing Corporation). 2011. *Urea-formaldehyde foam insulation* (UFFI). http://www.cmhc-schl.gc.ca/en/co/maho/yohoyohe/inaiqu/inaiqu_008.cfm (accessed May 3, 2011).

CPSC (US Consumer Product Safety Commission). 1982. *CPSC bans urea formaldehyde foam insulation (UFFI).* http://www.cpsc.gov/cpscpub/prerel/prhtml82/82005.html (accessed May 3, 2011).

CPSC. 1983. *Ban on UFFI lifted.* http://www.cpsc.gov/cpscpub/prerel/prhtml83/83048.html (accessed May 3, 2011).

Cobb N, Etzel RA. 1991. Unintentional carbon monoxide-related deaths in the United States, 1979 through 1988. *Journal of the American Medical Association* 266:659-663.

Cohen AJ, Gordis L. 1993. Introduction to the Health Effects Institute Environmental Epidemiology Planning Project documents. *Environmental Health Perspectives* 101(Suppl 4):15-17.

Coleman BK, Destaillats H, Hodgson AT, Nazaroff WW. 2008. Ozone consumption and volatile byproduct formation from surface reactions with aircraft cabin materials and clothing fabrics. *Atmospheric Environment* 42:642-654.

Colt JS, Davis S, Severson RK, Lynch CF, Cozen W, Camann D, Engels EA, Blair A, Hartge P. 2006. Residential insecticide use and risk of non-Hodgkin's lymphoma. *Cancer Epidemiology Biomarkers and Prevention* 15:251-257.

Dales R, Liu L, Wheeler AJ, Gilbert NL. 2008. Quality of indoor residential air and health. *Canadian Medical Association Journal* 179:147-152.

Daley WR, Shireley L, Gilmore R. 2001. A flood-related outbreak of carbon monoxide poisoning—Grand Forks, North Dakota. *Journal of Emergency Medicine* 21:249-253.

Dallman TR, Harley RA. 2010. Evaluation of mobile source emission trends in the United States. *Journal of Geophysical Research—Atmospheres* 115:Article No. D14305.

D'Amato G, Liccardi G, Frenguelli G. 2007. Thunderstorm asthma and pollen allergy. *Allergy* 62:11-16.

Darby S, Hill D, Auvinen A, Barrios-Dios JM, Baysson H, Bochicchio F, Deo H, Falk R, Forastiere F, Hakama M, Heid I, Kreienbrock L, Kreuzer M, Lagarde F, Mäkeläinen I, Muirhead C, Oberaigner W, Pershagen G, Ruano-Ravina A, Ruosteenoja E, Schaffrath Rosario A, Tirmarche M, Tomásek L, Whitley E, Wichmann HE, Doll R. 2005. Radon in homes and risk of lung cancer: Collaborative analysis of individual data from 13 European case-control studies. *British Medical Journal* 330:223-226.

DiFranza JR, Aligne CA, Weitzman W. 2004. Prenatal and postnatal environmental tobacco smoke exposure and children's health. *Pediatrics* 113:1007-1015.

Dockery DW, Pope CA, Xu XP, Spengler JD, Ware JH, Fay ME, Ferris BG, Speizer FE. 1993. An association between air pollution and mortality in 6 United States cities. *New England Journal of Medicine* 329:1753-1759.

Duty SM, Singh NP, Silva MJ, Barr DB, Brock JW, Ryan L, Herrick RF, Christiani DC, Hauser R. 2003. The relationship between environmental exposures to phthalates and DNA damage in human sperm using the neutral comet assay. *Environmental Health Perspectives* 111:1164-1169.

EPA (US Environmental Protection Agency). 2008. *The original list of hazardous air pollutants*. http://www.epa.gov/ttn/atw/188polls.html (accessed May 3, 2011).

EPA. 2009a. *Basic information*. http://www.epa.gov/air/emissions/basic.htm#dataloc (accessed February 17, 2011).

EPA. 2009b. *Pollutants & sources*. http://www.epa.gov/airtoxics/pollsour.html (accessed May 3, 2011).

EPA. 2010a. *Phthalates action plan summary*. http://www.epa.gov/opptintr/existingchemicals/pubs/actionplans/phthalates.html#action (accessed February 16, 2011).

EPA. 2010b. *Policy assessment for the review of the carbon monoxide national ambient air quality standards*. Research Triangle Park, North Carolina: Office of Air Quality Planning and Standards Health and Environmental Impacts Division.

EPA. 2010c. *The Clean Air Act Amendments of 1990 list of hazardous air pollutants*. http://www.epa.gov/ttn/atw/orig189.html (accessed February 16, 2011).

Erdmann CA, Apte MG. 2004. Mucous membrane and lower respiratory building related symptoms in relation to indoor carbon dioxide concentrations in the 100-building BASE dataset. *Indoor Air* 14(Suppl 8):127-134.

Eskenazi B, Bradman A, Castorina R. 1999. Exposures of children to organophosphate pesticides and their potential adverse health effects. *Environmental Health Perspectives* 107(Suppl 3):409-419.

Eskenazi B, Marks AR, Bradman A, Fenster L, Johnson C, Barr DB, Jewell NP. 2006. In utero exposure to dichlorodiphenyltrichloroethane (DDT) and dichlorodiphenyldichloroethylene (DDE) and neurodevelopment among young Mexican American children. *Pediatrics* 118:233-241.

Evans GJ, Peers A, Sabaliauskas K. 2008. Particle dose estimation from frying in residential settings. *Indoor Air* 18:499-510.

Ezzati M, Lopez AD, Rodgers A, Vander Hoorn S, Murray CJL. 2002. Selected major risk factors and global and regional burden of disease. *Lancet* 360:1347-1360.

Fine PM, Cass GR, Simoneit BRT. 1999. Characterization of fine particle emissions from burning church candles. *Environmental Science & Technology* 33:2352-2362.

Fischer D, Hooper K, Athanasiadou M, Athanassiadis I, Bergman Å. 2006. Children show highest levels of polybrominated diphenyl ethers in a California family of four: A case study. *Environmental Health Perspectives* 114(10):1581-1584.

Fisk WJ, Faulkner D, Palonen J, Seppänen O. 2002. Performance and costs of particle air filtration technologies. *Indoor Air* 12:223-234.

Fleming LE, Kirkpatrick B, Backer LC, Bean JA, Wanner A, Dalpra D, Tamer R, Zaias J, Cheng YS, Pierce R, Naar J, Abraham W, Clark R, Zhou Y, Henry MS, Johnson D, Van de Bogart G, Bossart GD, Harrington M, Baden DG. 2005. Initial evaluation of the effects of aerosolized Florida red tide toxins (brevetoxins) in persons with asthma. *Environmental Health Perspectives* 113:650-657.

Fleming LE, Bean JA, Kirkpatrick B, Cheng YS, Pierce R, Naar J, Nierenberg K, Backer LC, Wanner A, Reich A, Zhou Y, Watkins S, Henry M, Zaias J, Abraham WM, Benson J, Cassedy A, Hollenbeck J, Kirkpatrick G, Clarke T, Baden DG. 2009. Exposure and effect assessment of aerosolized red tide toxins (brevetoxins) and asthma. *Environmental Health Perspectives* 117:1095-1100.

Fontham ETH, Correa P, Reynolds P, Wu-Williams A, Buffler PA, Greenberg RS, Chen VW, Alterman T, Boyd P, Austin DF, Liff J. 1994. Environmental tobacco smoke and lung cancer in nonsmoking women—A multicenter study. *Journal of the American Medical Association* 271:1752-1759.

Forrester MB. 2009. Impact of Hurricane Ike on Texas Poison Center calls. *Disaster Medicine and Public Health Preparedness* 3(3):151-157.

Francisco PW, Gordon JR, Rose B. 2010. Measured concentrations of combustion gases from the use of unvented gas fireplaces. *Indoor Air* 20(5):370-379.

Galada HC, Gurian PL, Corella-Barud V, Pérez FG, Velázquez-Angulo G, Flores S, Montoya T. 2009. Applying the mental models framework to carbon monoxide risk in northern Mexico. *Revista Panamericana de salud Pública/Pan American Journal of Public Health* 25(3):242-253.

Garrett MH, Hooper MA, Hooper BM, Rayment PR, Abramson MJ. 1999. Increased risk of allergy in children due to formaldehyde exposure in homes. *Allergy* 54(4):330-337.

George K, Ziska LH, Bunce JA, Quebedeaux B. 2007. Elevated atmospheric CO_2 concentration and temperature across an urban-rural transect. *Atmospheric Environment* 41:7654-7665.

Gilbert NL, Gauvin D, Guay M, Héroux ME, Dupuis G, Legris M, Chan CC, Dietz RN, Lévesque B. 2006. Housing characteristics and indoor concentrations of nitrogen dioxide and formaldehyde in Québec City, Canada. *Environment Research* 102:1-8.

Ginevan ME, Mills WA. 1986. Assessing the risks of radon exposure: The influence of cigarette smoking. *Health Physics* 51:163-174.

Girman JR, Chang YL, Hayward SB, Liu KS. 1998. Causes of unintentional deaths from carbon monoxide poisonings in California. *Western Journal of Medicine* 168:158-165.

Glibert PM, Anderson DM, Gentien P, Granéli E, Sellner KG. 2005. The global, complex phenomena of harmful algal blooms. *Oceanography* 18(2):137-147.

Grøntoft T, Raychaudhuri MR. 2004. Compilation of tables of surface deposition velocities for O_3, NO_2 and SO_2 to a range of indoor surfaces. *Atmospheric Environment* 38:533-544.

Gustafson P, Östman C, Sällsten G. 2008. Indoor levels of polycyclic aromatic hydrocarbons in homes with or without wood burning for heating. *Environmental Science & Technology* 42:5074-5080.

Hampson NB, Stock AL. 2006. Storm-related carbon monoxide poisoning: Lessons learned from recent epidemics. *Undersea & Hyperbaric Medicine* 33(4):257-263.

Hanley JT, Ensor DS, Smith DD, Sparks LE. 1994. Fractional aerosol filtration efficiency of in-duct ventilation air cleaners. *Indoor Air* 4:169-178.

Harduar-Morano L, Watkins S. 2011. Review of unintentional non-fire-related carbon monoxide poisoning morbidity and mortality in Florida, 1999-2007. *Public Health Reports* 126:240-250.

Harley KG, Marks A, Chevrier J, Bradman A, Sjödin A, Eskenazi B. 2010. PBDE concentrations in women's serum and fecundability. *Environmental Health Perspectives* 118(5):699-704.

Hauser R, Calafat AM. 2005. Phthalates and human health. *Occupational and Environmental Medicine* 62:806-818.

Haverinen-Shaughnessy U, Moschandreas DJ, Shaughnessy RJ. 2011. Association between sub-standard classroom ventilation rates and students' academic achievement. *Indoor Air* 21(2):121-131.

Heath GA, Granvold PW, Hoats AS, Nazaroff WW. 2006. Intake fraction assessment of the air pollutant exposure implications of a shift toward distributed electricity generation. *Atmospheric Environment* 40:7164-7177.

Heath GA, Nazaroff WW. 2007. Intake to delivered energy ratios for central station and distributed electricity generation in California. *Atmospheric Environment* 41:9159-9172.

Hefflin BJ, Jalaludin B, McClure E, Cobb N, Johnson CA, Jecha L, Etzel RA. 1994. Surveillance for dust storms and respiratory diseases in Washington State, 1991. *Archives of Environmental Health* 49:170-174.

Heinzow B, Sagunski H. 2009. Evaluation of indoor air contamination by means of reference and guide values: the German approach. In *Organic indoor air pollutants: Occurrence, measurement, evaluation* edited by Salthammer T, Uhde, E. Federal Republic of Germany: Wiley-VCH.

Herbarth O, Matysik S. 2010. Decreasing concentrations of volatile organic compounds (VOC) emitted following home renovations. *Indoor Air* 20:141-146.

Herbstman JB, Sjödin A, Kurzon M, Lederman SA, Jones RS, Rauh V, Needham LL, Tang D, Niedzwiecki M, Wang R, Perera F. 2010. Prenatal exposure to PBDEs and neurodevelopment. *Environmental Health Perspectives* 118(5):712-719.

Herrick RF, McClean MD, Meeker JD, Baxter LK, Weymouth GA. 2004. An unrecognized source of PCB contamination in schools and other buildings. *Environmental Health Perspectives* 112:1051-1053.

Hnatov MV. 2009. *Non-fire carbon monoxide deaths associated with the use of consumer products—2006 annual estimates.* http://www.cpsc.gov/library/co09.pdf (accessed October 15, 2010).

Hogrefe C, Biswas J, Lynn B, Civerolo K, Ku JY, Rosenthal J, Rosenzweig C, Goldberg R, Kinney PL. 2004a. Simulating regional-scale ozone climatology over the eastern United States: Model evaluation results. *Atmospheric Environment* 38(17):2627-2638.

Hogrefe C, Lynn B, Civerolo K, Ku JY, Rosenthal J, Rosenzweig C, Goldberg R, Gaffin S, Knowlton K, Kinney PL. 2004b. Simulating changes in regional air pollution over the eastern United States due to changes in global and regional climate and emissions. *Journal of Geophysical Research* 109:D22301.

Holmes KJ, Russell AG. 2004. Why carbon monoxide still matters. *Environmental Science & Technology* 38:288A-294A.

Hwang HM, Park EK, Young TM, Hammock BD. 2008. Occurrence of endocrine-disrupting chemicals in indoor dust. *Science of the Total Environment* 404(1):26-35.

Hyland A, Travers MJ, Dresler C, Higbee C, Cummings KM. 2008. A 32-county comparison of tobacco smoke derived particle levels in indoor public places. *Tobacco Control* 17:159-165.

Hystad PU, Setton EM, Allen RW, Keller PC, Brauer M. 2009. Modeling residential fine particulate matter infiltration for exposure assessment. *Journal of Exposure Science and Environmental Epidemiology* 19:570-579.

IARC (International Agency for Research on Cancer). 2006. *IARC Monographs on the evaluation of carcinogenic risks to humans, volume 88, formaldehyde, 2-butoxyethanol and 1-tert-butoxypropan-2-ol.* Geneva: WHO Press.

IEA (International Energy Agency)/UNDP (United Nations Development Programme)/UNIDO (United Nations Industrial Development Organization). 2010. *Energy poverty: How to make world energy access universal.* http://www.worldenergyoutlook.org/docs/weo2010/weo2010_poverty.pdf (accessed January 7, 2011).

IOM (Institute of Medicine). 1981. *Indoor pollutants.* Washington, DC: National Academy Press.

IOM. 2000. *Clearing the air: Asthma and indoor air exposures.* Washington, DC: National Academy Press.

IOM. 2008. *Global climate change and extreme weather events: Understanding the contributions to infectious disease emergence: Workshop summary.* Washington, DC: The National Academies Press.

Iqbal S, Clower JH, Boehmer TK, Yip FY, Garbe P. 2010. Carbon monoxide-related hospitalizations in the US: Evaluation of a web-based query system for public health surveillance. *Public Health Reports* 125:423-432.

Jacob DJ, Winner DA. 2009. Effect of climate change on air quality. *Atmospheric Environment* 43:51-63.

Jacobson JL, Jacobson SW, Humphrey HE. 1990. Effects of in utero exposure to polychlorinated biphenyls and related contaminants on cognitive functioning in young children. *Journal of Pediatrics* 116:38-45.

Jacobson JL, Jacobson SW. 1996. Intellectual impairment in children exposed to polychlorinated biphenyls in utero. *New England Journal of Medicine* 335:783-789.

Jarvis D, Chinn S, Luczynska C, Burney P. 1996. Association of respiratory symptoms and lung function in young adults with use of domestic gas appliances. *Lancet* 347:426-431.

Jerrett M, Burnett RT, Pope CA, Ito K, Thurston G, Krewski D, Shi YL, Calle E, Thun M. 2009. Long-term ozone exposure and mortality. *New England Journal of Medicine* 360:1085-1095.

Jetter JJ, Guo ZS, McBrian JA, Flynn MR. 2002. Characterization of emissions from burning incense. *Science of the Total Environment* 295:61-67.

Jia C, Batterman S, Godwin C. 2008a. VOCs in industrial, urban and suburban neightborhoods, Part 1: Indoor and outdoor concentrations, variation, and risk drivers. *Atmospheric Environment* 42:2083-2100.

Jia C, Batterman S, Godwin C. 2008b. VOCs in industrial, urban and suburban neighborhoods, Part 2: Factors affecting indoor and outdoor concentrations. *Atmospheric Environment* 42:2101-2116.

Jones AP. 1999. Indoor air quality and health. *Atmospheric Environment* 33:4535-4564.

Kattan M, Gergen PJ, Eggleston P, Visness CM, Mitchell HE. 2007. Health effects of indoor nitrogen dioxide and passive smoking on urban asthmatic children. *Journal of Allergy and Clinical Immunology* 120:618-624.

Keeling RF, Piper SC, Bollenbacher AF, Walker JS. 2009. Atmospheric CO_2 records from sites in the SIO air sampling network. In *Trends: A compendium of data on global change.* Oak Ridge, TN: Carbon Dioxide Information Analysis Center, Oak Ridge National Laboratory, US Department of Energy.

King M, Bailey C. 2008. Carbon monoxide related deaths United States, 1999-2004. *Journal of the American Medical Association* 299:1011-1012.

Kinney PL. 2008. Climate change, air quality, and human health. *American Journal of Preventive Medicine* 35:459-467.

Kirkpatrick B, Fleming LE, Backer LC, Bean JA, Tamer R, Kirkpatrick G, Kane T, Wanner A, Dalpra D, Reich A, Baden DG. 2006. Environmental exposures to Florida red tides: Effects on emergency room respiratory diagnoses admissions. *Harmful Algae* 5:526-533.

Kittelson DB, 1998. Engines and nanoparticles; A review. *Journal of Aerosol Science* 29:575-588.

Klein KR, Herzog P, Smolinske S, White SR. 2007. Demand for poison control center services "surged" during the 2003 blackout. *Clinical Toxicology* 45(3):248-254.

Ko YC, Lee CH, Chen MJ, Huang CC, Chang WY, Lin HJ, Wang HZ, Chang PY. 1997. Risk factors for primary lung cancer among non-smoking women in Taiwan. *International Journal of Epidemiology* 26:24-31.

Kraev TA, Adamkiewicz G, Hammond SK, Spengler JD. 2009. Indoor concentrations of nicotine in low-income, multi-unit housing: Associations with smoking behaviours and housing characteristics. *Tobacco Control* 18(6):438-44.

Krewski D, Lubin JH, Zielinski JM, Alavanja M, Catalan VS, Field RW, Klotz JB, Létourneau EG, Lynch CF, Lyon JI, Sandler DP, Schoenberg JB, Steck DJ, Stolwijk JA, Weinberg C, Wilcox HB. 2005. Residential radon and risk of lung cancer: A combined analysis of 7 North American case-control studies. *Epidemiology* 16:137-145.

Kulle TJ. 1993. Acute odor and irritation response in healthy nonsmokers with formaldehyde exposure. *Inhalation Toxicology* 5:323-332.

Kulmala M, Vehkamäki H, Petäjä T, Dal Maso M, Lauri A, Kerminen V-M, Birmili W, McMurry PH. 2004. Formation and growth rates of ultrafine atmospheric particles: A review of observations. *Journal of Aerosol Science* 35:143-176.

Kunkel DA, Gall ET, Siegel JA, Novoselac A, Morrison GC, Corsi RL. 2010. Passive reduction of human exposure to indoor ozone. *Building and Environment* 45:445-452.

Kuo HW, Shen HY. 2010. Indoor and outdoor PM2.5 and PM10 concentrations in the air during a dust storm. *Building and Environment* 45:610-614.

L'Abbé KA, Hoey JR. 1984. Review of the health effects of urea-formaldehyde foam insulation. *Environmental Research* 35:246-263.

Lai ACK, Nazaroff WW. 2000. Modeling indoor particle deposition from turbulent flow onto smooth surfaces. *Journal of Aerosol Science* 31:463-476.

Leaderer BP. 1982. Air pollutant emissions from kerosene space heaters. *Science* 218:1113-1115.

Leaderer BP, Boone PM, Hammond SK. 1990. Total particle, sulfate, and acidic aerosol emissions from kerosene space heaters. *Environmental Science & Technology* 24:908-912.

Lee P, Davidson J. 1999. Evaluation of activated carbon filters for removal of ozone at the ppb level. *American Industrial Hygiene Association Journal* 60:589-600.

Lenes JM, Darrow BP, Cattrall C, Heil CA, Callahan M, Vargo GA, Byrne RH, Prospero JM, Bates DE, Fanning KA, Walsh JJ. 2001. Iron fertilization and the *Trichodesmium* response on the West Florida shelf. *Limnology and Oceanography* 46:1261-1277.

Lightwood JM, Glantz SA. 2009. Declines in acute myocardial infarction after smoke-free laws and individual risks attributable to secondhand smoke. *Circulation* 120:1373-1379.

Liu DL, Nazaroff WW. 2001. Modeling pollutant penetration across building envelopes. *Atmospheric Environment* 35:4451-4462.

Liu KS, Paz MK, Flessel P, Waldman J, Girman J. 2000. Unintentional carbon monoxide deaths in California from residential and other nonvehicular sources. *Archives of Environmental Health* 55:375-381.

Liu W, Zhang J, Hashim JH, Jalaludin J, Hashim Z, Goldstein BD. 2003. Mosquito coil emissions and health implications. *Environmental Health Perspectives* 111:1454-1460.

Logue JM, McKone TE, Sherman MH, Singer BC. 2011. Hazard assessment of chemical air contaminants measured in residences, *Indoor Air* 21(2):92-109.

Lopez AD, Mathers CD, Ezzati M, Jamison DT, Murray CJL. 2006. Global and regional burden of disease and risk factors, 2001: Systematic analysis of population health data. *Lancet* 367:1747-1757.

Lunden MM, Revzan KL, Fischer ML, Thatcher TL, Littlejohn D, Hering SV, Brown NJ. 2003. The transformation of outdoor ammonium nitrate aerosols in the indoor environment. *Atmospheric Environment* 37:5633-5644.

Maddalena R, Russell M, Sullivan DP, Apte MG. 2009. Formaldehyde and other volatile organic chemical emissions in four FEMA temporary housing units. *Environmental Science & Technology* 43(15):5626-5632.

Malig BJ, Ostro BD. 2009. Coarse particles and mortality: Evidence from a multi-city study in California. *Occupational and Environmental Medicine* 66:832-839.

Manzey D, Lorenz B. 1998. Joint NASA-ESA-DARA Study—Part three: Effects of chronically elevated CO_2 on mental performance during 26 days of confinement. *Aviation Space and Environmental Medicine* 69:506-514.

Marbury MC, Harlos DP, Samet JM, Spengler JD. 1988. Indoor residential NO_2 concentrations in Albuquerque, New Mexico. *The International Journal of Air Pollution Control and Hazardous Waste Management* 38:392-398.

Marcinowski F, Lucas RM, Yeager WM. 1994. National and regional distributions of airborne radon concentrations in U.S. homes. *Health Physics* 66(6):699-706.

Margel D, White DP, Pillar G. 2003. Long-term intermittent exposure to high ambient CO_2 causes respiratory disturbances during sleep in submariners. *Chest* 124:1716-1723.

McCormack MC, Breysse PN, Hansel NN, Matsui EC, Tonorezos ES, Curtin-Brosnan J, Williams DL, Buckley TJ, Eggleston PA, Diette GB. 2008. Common household activities are associated with elevated particulate matter concentrations in bedrooms of inner-city Baltimore pre-school children. *Environmental Research* 106:148-155.

McDonald JD, Zielinska B, Fujita EM, Sagebiel JC, Chow JC, Watson JG. 2000. Fine particle and gaseous emission rates from residential wood combustion. *Environmental Science & Technology* 34:2080-2091.

Mendell MJ. 2007. Indoor residential chemical emissions as risk factors for respiratory and allergic effects in children: A review. *Indoor Air* 17:259-277.

Milian A, Nierenberg K, Fleming LE, Bean JA, Wanner A, Reich A, Backer LC, Jayroe D, Kirkpatrick B. 2007. Reported respiratory symptom intensity in asthmatics during exposure to aerosolized Florida red tide toxins. *Journal of Asthma* 44:583-587.

Mills E. 2005. The specter of fuel-based lighting. *Science* 308:1263-1264.

Mishra VK, Retherford RD, Smith KR. 1999. Biomass cooking fuels and prevalence of tuberculosis in India. *International Journal of Infectious Diseases* 3:119-129.

Mohan M, Sperduto RD, Angra SK, Milton RC, Mathur RL, Underwood BA, Jaffery N, Pandya CB, Chhabra VK, Vajpayee RB, Kalra VK, Sharma YR. 1989. India–United States case-control study of age-related cataracts. *Archives of Ophthalmology* 107:670-676.

Monn C. 2001. Exposure assessment of air pollutants: A review on spatial heterogeneity and indoor/outdoor/personal exposure to suspended particulate matter, nitrogen dioxide and ozone. *Atmospheric Environment* 35:1-32.

Morales E, Julvez J, Torrent M, de Cid R, Guxens M, Bustamante M, Kunzli N, Sunyer J. 2009. Association of early-life exposure to household gas appliances and indoor nitrogen dioxide with cognition and attention behavior in preschoolers. *American Journal of Epidemiology* 169:1327-1336.

Mott JA, Wolfe MI, Alverson CJ, MacDonald SC, Bailey CR, Ball LB, Moorman JE, Somers JH, Mannino DM, Redd SC. 2002. National vehicle emissions policies and practices and declining US carbon monoxide-related mortality. *Journal of the American Medical Association* 288:988-995.

Mullen NA, Liu C, Zhang Y, Wang S, Nazaroff WW. 2010. Ultrafine particle concentrations and exposures in four high-rise Beijing apartments. *Atmospheric Environment*, accepted for publication, doi:10.1016/j.atmosenv.2010.07.060.

Mumford JL, Li XM, Hu FD, Lu XB, Chuang JC. 1995. Human exposure and dosimetry of polycyclic aromatic hydrocarbons in urine from Xuan Wei, China with high lung cancer mortality associated with exposure to unvented coal smoke. *Carcinogenesis* 16:3031-3036.

Muscatiello NA, Babcock G, Jones R, Horn E, Hwang S-A. 2010. Hospital emergency department visits for carbon monoxide poisoning following an October 2006 snowstorm in western New York. *Journal of Environmental Health* 72(6):43-48.

Naeher LP, Brauer M, Lipsett M, Zelikoff JT, Simpson CD, Koenig JQ, Smith KR. 2007. Woodsmoke health effects: A review. *Inhalation Toxicology* 19:67-106.

Nagda NL, Koontz MD, Billick IH, Leslie NP, Behrens DW. 1996. Causes and consequences of backdrafting of vented gas appliances. *Journal of the Air & Waste Management Association* 46:838-846.

National Candle Association. 2011. *Facts & figures*. http://www.candles.org/about_facts.html (accessed February 16, 2011).

National Conference of State Legislatures. 2010. *Carbon monoxide detectors state statutes*. http://www.ncsl.org/IssuesResearch/EnvironmentandNaturalResources/CarbonMonoxideDetectorsStateStatutes/tabid/13238/Default.aspx (accessed October 13, 2010).

Nazaroff WW. 1992. Radon transport from soil to air. *Reviews of Geophysics* 30:137-160.

Nazaroff WW. 2004. Indoor particle dynamics. *Indoor Air* 14(Suppl 7):175-183.

Nazaroff WW, Klepeis NE. 2004. Environmental tobacco smoke particles. In *Indoor environment: Airborne particles and settled dust*, edited by Morawska L, Salthammer T. Weinheim: Wiley-VCH Verlag GmbH & Co. KGaA.

Nazaroff WW, Singer BC. 2004. Inhalation of hazardous air pollutants from environmental tobacco smoke in US residences. *Journal of Exposure Analysis and Environmental Epidemiology* 14:S71-S77.

Nazaroff WW. 2008. Inhalation intake fraction of pollutants from episodic indoor emissions. *Building and Environment* 43:269-277.

Neas LM, Dockery DW, Ware JH, Spengler JD, Speizer FE, Ferris BG. 1991. Association of indoor nitrogen dioxide with respiratory symptoms and pulmonary function in children. *American Journal of Epidemiology* 134:204-219.

Norbäck D, Wieslander G, Nordström K, Walinder R. 2000. Asthma symptoms in relation to measured building dampness in upper concrete floor construction, and 2-ethyl-1-hexcanol in indoor air. *The International Journal of Tuberculosis and Lung Disease* 4:1016-1025.

NRC (National Research Council). 1981. *Indoor pollutants*. Washington, DC: National Academy Press.

NRC. 2010. *Advancing the science of climate change*. Washington, DC: The National Academies Press.

Oberdörster G. 2001. Pulmonary effects of inhaled ultrafine particles. *International Archives of Occupational and Environmental Health* 74:1-8.

OEHHA (Office of Environmental Health Hazard Assessment). 2007. *Air toxicology and epidemiology*. http://oehha.ca.gov/air/allrels.html (accessed May 3, 2011).

Offermann FJ. 2009. *Ventilation and indoor air quality in new homes*. Collaborative Report. CEC-500-2009-085. PIER Energy-Related Environmental Research Program. Sacramento, CA: California Air Resources Board and California Energy Commission.

Olson DA, Burke JM. 2006. Distributions of PM2.5 source strengths for cooking from the Research Triangle Park Particulate Matter Panel study. *Environmental Science & Technology* 40:163-169.

Olsson M, Kjallstrand J. 2006. Low emissions from wood burning in an ecolabelled residential boiler. *Atmospheric Environment* 40:1148-1158.

Ostro BD, Broadwin R, Lipsett MJ. 2000. Coarse and fine particles and daily mortality in the Coachella Valley, California: A follow-up study. *Journal of Exposure Analysis and Environmental Epidemiology* 10:412-419.

Pagels J, Wierzbicka A, Nilsson E, Isaxon C, Dahl A, Gudmundsson A, Swietlicki E, Bohgard M. 2009. Chemical composition and mass emission factors of candle smoke particles. *Journal of Aerosol Science* 40:193-208.

Pandrangi LS, Morrison GC. 2008. Ozone interactions with human hair: Ozone uptake rates and product formation. *Atmospheric Environment* 42:5079-5089.

Park RJ, Jacob DJ, Logan JA. 2007. Fire and biofuel contributions to annual mean aerosol mass concentrations in the United States. *Atmospheric Environment* 41:7389-7400.

Paustenbach D, Galbraith D. 2006. Biomonitoring and biomarkers: Exposure assessment will never be the same. *Environmental Health Perspectives* 114:1143-1149.

Pawel DJ, Puskin JS. 2004. The US Environmental Protection Agency's assessment of risks from indoor radon. *Health Physics* 87:68-74.

Pepermans G, Driesen J, Haeseldonckx D, Belmans R, D'haeseleer W. 2005. Distributed generation: Definition, benefits and issues. *Energy Policy* 33:787-798.

Perera FP, Rauh V, Tsai WY, Kinney P, Camann D, Barr D, Bernert T, Garfinkel R, Tu YH, Diaz D, Dietrich J, Whyatt RM. 2003. Effects of transplacental exposure to environmental pollutants on birth outcomes in a multiethnic population. *Environmental Health Perspectives* 111:201-205.

Pintos J, Franco EL, Kowalski LP, Oliveira BV, Curado MP. 1998. Use of wood stoves and risk of cancers of the upper aero-digestive tract: A case-control study. *International Journal of Epidemiology* 27:936-940.

Pirkle JL, Bernert JT, Caudill SP, Sosnoff CS, Pechacek TF. 2006. Trends in the exposure of nonsmokers in the U.S. population to secondhand smoke: 1988-2002. *Environmental Health Perspectives* 114:853-856.

Pope CA III, Dockery DW. 2006. Health effects of fine particulate air pollution: Lines that connect. *Journal of the Air & Waste Management Association* 56:709-742.

Pope CA III, Ezzati M, Dockery DW. 2009. Fine-particulate air pollution and life expectancy in the United States. *New England Journal of Medicine* 360:376-386.

Porstendörfer J. 1994. Properties and behavior of radon and thoron and their decay products in air. *Journal of Aerosol Science* 25:219-263.

Quackenboss JJ, Spengler JD, Kanarek MS, Letz R, Duffy CP. 1986. Personal exposure to nitrogen dioxide: Relationship to indoor/outdoor air quality and activity patterns. *Environmental Science & Technology* 20:775-783.

Racheria PN, Adams PJ. 2009. US ozone air quality under changing climate and anthropogenic emissions. *Environmental Science & Technology* 43:571-577.

Rahman NM, Tracy BL. 2009. Radon control systems in existing and new construction: A review. *Radiation Protection Dosimetry* 135:243-255.

Ramanathan V, Carmichael G. 2008. Global and regional climate changes due to black carbon. *Nature Geoscience* 1:221-227.

Reid CE, Gamble JL. 2009. Aeroallergens, allergic disease and climate change: Impacts and adaptation. *Ecohealth* 6:458-470.

Riley WJ, McKone TE, Lai ACK, Nazaroff WW. 2002. Indoor particulate matter of outdoor origin: Importance of size-dependent removal mechanisms. *Environmental Science & Technology* 36:200-207. (See also errata p. 1868.)

Rogan WJ, Gladen BC, McKinney JD, Carreras N, Hardy P, Thullen J, Tinglestad J, Tully M. 1986. Neonatal effects of transplacental exposure to PCBs and DDE. *Journal of Pediatrics* 109:335-341.

Rogan WJ, Gladen BC. 1991. PCBs, DDE, and child development at 18 and 24 months. *Annals of Epidemiology* 1:407-413.

Roosens L, Abdallah MAE, Harrad S, Neels H, Covaci A. 2009. Exposure to hexabromocyclododecanes (HBCDs) via dust ingestion, but not diet, correlates with concentrations in human serum: Preliminary results. *Environmental Health Perspectives* 117:1707-1712.

Rosas LG, Eskenazi B. 2008. Pesticides and child neurodevelopment. *Current Opinion in Pediatrics* 20(2):191-197.

Rudel RA, Camann DE, Spengler JD, Korn LR, Brody JG. 2003. Phthalates, alkylphenols, pesticides, polybrominated diphenyl ethers, and other endocrine-disrupting compounds in indoor air and dust. *Environmental Science & Technology* 37:4543-4553.

Rudel RA, Seryak LM, Brody JG. 2008. PCB-containing wood floor finish is a likely source of elevated PCBs in residents' blood, household air and dust: A case study of exposure. *Environmental Health* 7:Article 2.

Rudel RA, Perovich LJ. 2009. Endocrine disrupting chemicals in indoor and outdoor air. *Atmospheric Environment* 43:170-181.

Rudel RA, Dodson RE, Perovich LJ, Morello-Frosch R, Camann DE, Zuniga MM, Yau AY, Just AC, Brody JG. 2010. Semivolatile endocrine-disrupting compounds in paired indoor and outdoor air in two northern California communities. *Environmental Science & Technology* 44:6583-6590.

Ruiz PA, Toro C, Cáceres J, López G, Ovola P, Koutrakis P. 2010. Effect of gas and kerosene space heaters on indoor air quality: A study in homes of Santiago, Chile. *Journal of Air & Waste Management Association* 60(1):98-108.

Runyan CW, Johnson RM, Yang J, Waller AE, Perkis D, Marshall SW, Coyne-Beasley T, McGee KS. 2005. Risk and protective factors for fires, burns, and carbon monoxide poisoning in U.S. households. *American Journal of Preventative Medicine* 28(1):102-108.

Russell M, Sherman M, Rudd A. 2007. Review of residential ventilation technologies. *HVAC&R Research* 13:325-348.

Salthammer T, Fuhrmann F, Kaufhold S, Meyer B, Schwarz A. 1995. Effects of climatic parameters on formaldehyde concentrations in indoor air. *Indoor Air* 5:120-128.

Salthammer T, Bahadir M. 2009. Occurrence, dynamics, and reactions of organic pollutants in the indoor environment. *Clean* 37:417-435.

Salthammer T, Mentese S, Marutzky R. 2010. Formaldehyde in the indoor environment. *Chemical Reviews* 110:2536-2572.

Samet JM, Marbury MC, Spengler JD. 1987. Health effects and sources of indoor air pollution. 1. *American Review of Respiratory Disease* 136:1486-1508.

Samet JM, Marbury MC, Spengler JD. 1988. Health effects and sources of indoor air pollution. 2. *American Review of Respiratory Disease* 137:221-242.

Samet JM, Lambert WE, Skipper BJ, Cushing AH, Hunt WC, Young SA, McLaren LC, Schwab M, Spengler JD. 1993. Nitrogen dioxide and respiratory illnesses in infants. *American Review of Respiratory Disease* 148:1258-1265.

Seppänen OA, Fisk WJ, Mendell MJ. 1999. Association of ventilation rates and CO_2 concentrations with health and other responses in commercial and institutional buildings. *Indoor Air* 9:226-252.

Sexton K, Adgate JL, Fredrickson AL, Ryan AD, Needham LL, Ashley DL. 2006. Using biologic markers in blood to assess exposure to multiple environmental chemicals for inner-city children 3–6 years of age. *Environmental Health Perspectives* 114:453-459.

Shair FH. 1981. Relating indoor pollutant concentrations of ozone and sulfur dioxide to those outside: Economic reduction of indoor ozone through selective filtration of the make-up air. *ASHRAE Transactions* 87(Part 1):116-139.

Shendell DG, Prill R, Fisk WJ, Apte MG, Blake D, Faulkner D. 2004. Associations between classroom CO_2 concentrations and student attendance in Washington and Idaho. *Indoor Air* 14:333-341.

Singh GK, Siahpush M, Kogan MD. 2010. Disparities in children's exposure to environmental tobacco smoke in the United States, 2007. *Pediatrics* 126:4-13.

Sioutas C, Delfino RJ, Sing M. 2005. Exposure assessment for atmospheric ultrafine particles (UFPs) and implications in epidemiologic research. *Environmental Health Perspectives* 113:947-955.

Sippola MR, Nazaroff WW. 2003. Modeling particle loss in ventilation ducts. *Atmospheric Environment* 37:5597-5609.

Slama K, Chiang C-Y, Hinderaker SG, Bruce N, Vedal S, Enarson DA. 2010. Indoor solid fuel combustion and tuberculosis: is there an association? *The International Journal of Tuberculosis and Lung Disease* 14(1):6-14.

Smedje G, Norback D, Edling C. 1997. Asthma among secondary schoolchildren in relation to the school environment. *Clinical & Experimental Allergy* 27(11):1270-1278.

Smith KR. 1988. Air pollution: Assessing total exposure in the United States. *Environment* 30(8):10-15, 33-38.

Smith KR. 1993. Fuel combustion, air pollution exposure, and health: The situation in developing countries. *Annual Review of Energy and the Environment* 18:529-566.

Smith KR. 2000. National burden of disease in India from indoor air pollution. *Proceedings of the National Academy of Sciences of the United States of America* 97(24):13286-13293.

Spengler J, Schwab M, Ryan PB, Colome S, Wilson AL, Billick I, Becker E. 1994. Personal exposure to nitrogen dioxide in the Los Angeles basin. *Journal of the Air & Waste Management Association* 44:39-47.

Spracklen DV, Mickley LJ, Logan JA, Hudman RC, Yevich R, Flannigan MD, Westerling AL. 2009. Impacts of climate change from 2000 to 2050 on wildfire activity and carbonaceous aerosol concentrations in the western United States. *Journal of Geophysical Research—Atmospheres* 114:Article D20301.

Tagaris E, Manomaiphiboon K, Liao KJ, Leung LR, Woo JH, He S, Amar P, Russell AG. 2007. Impacts of global climate change and emissions on regional ozone and fine particulate matter concentrations over the United States. *Journal of Geophysical Research—Atmospheres* 112:Article No. D14312.

Tagaris E, Liao KJ, Delucia AJ, Deck L, Amar P, Russell AG. 2009. Potential impact of climate change on air pollution related human health effects. *Environmental Science & Technology* 43:4979-4988.

Ten Brinke J, Selvin S, Hodgson AT, Fisk WJ, Mendell MJ, Koshland CP, Daisey JM. 1998. Development of new volatile organic compound (VOC) exposure metrics and their relationship to "sick building syndrome" symptoms. *Indoor Air* 8:140-152.

Thatcher TL, Lai ACK, Moreno-Jackson R, Sextro RG, Nazaroff WW. 2002. Effects of room furnishings and air speed on particle deposition rates indoors. *Atmospheric Environment* 36:1811-1819.

Traynor GW, Apte MG, Carruthers AR, Dillworth JF, Grimsrud DT, Gundel LA. 1987. Indoor air pollution due to emissions from wood-burning stoves. *Environmental Science & Technology* 21:691-697.

Triche EW, Belanger K, Bracken MB, Beckett WS, Holford TR, Gent JF, McSharry JE, Leaderer BP. 2005. Indoor heating sources and respiratory symptoms in nonsmoking women. *Epidemiology* 16:377-384.

US Department of Labor. 1989. *Final rules 54:2332-2983*. http://www.osha.gov/pls/oshaweb/owadisp.show_document?p_table=FEDERAL_REGISTER&p_id=12908 (accessed February 17, 2011).

US Department of Labor. 1990. *Carbon dioxide in workplace atmospheres*. http://www.osha.gov/dts/sltc/methods/inorganic/id172/id172.html (accessed February 17, 2011).

US Department of Transportation. 2004. *Census 2000 population statistics: US population living in urban vs. rural areas*. http://www.fhwa.dot.gov/planning/census/cps2k.htm (accessed February 16, 2011).

Valero N, Aguilera I, Llop S, Esplugues A, de Nazelle A, Ballester F, Sunyer J. 2009. Concentrations and determinants of outdoor, indoor and personal nitrogen dioxide in pregnant women from two Spanish birth cohorts. *Environment International* 35:1196-1201.

Van Sickle D, Chertow DS, Schulte JM, Ferdinands JM, Patel PS, Johnson DR, Harduar-Morano L, Blackmore C, Ourso AC, Cruse KM, Dunn KH, Moolenaar RL. 2007. Carbon monoxide poisoning in Florida during the 2004 hurricane season. *American Journal of Preventive Medicine* 32:340-346.

Van Strien RT, Gent JF, Belanger K, Triche E, Bracken MB, Leaderer BP. 2004. Exposure to NO_2 and nitrous acid and respiratory symptoms in the first year of life. *Epidemiology* 15:471-478.

Wallace L. 1996. Indoor particles: A review. *Journal of the Air & Waste Management Association* 46:98-126.

Wallace L, Emmerich SJ, Howard-Reed C. 2004. Source strengths of ultrafine and fine particles due to cooking with a gas stove. *Environmental Science & Technology* 38:2304-2311.

Wallace L, Wang F, Howard-Reed C, Persily A. 2008. Contribution of gas and electric stoves to residential ultrafine particle concentrations between 2 and 64 nm: Size distributions and emission and coagulation rates. *Environmental Science & Technology* 42:8641-8647.

Wallace LA, Pellizzari ED, Hartwell TD, Sparacino CM, Sheldon LS, Zelon H. 1985. Personal exposures, indoor-outdoor relationships, and breath levels of toxic air pollutants measured for 355 persons in New Jersey. *Atmospheric Environment* 19:1651-1661.

Wallace LA, Pellizzari ED, Hartwell TD, Sparacino C, Whitmore R, Sheldon L, Zelon H, Perritt R. 1987. The TEAM study: Personal exposures to toxic substances in air, drinking water, and breath of 400 residents of New Jersey, North Carolina, and North Dakota. *Environmental Research* 43:290-307.

Wallace LA, Pellizzari ED, Hartwell TD, Whitmore R, Zelon H, Perritt R, Sheldon L. 1988. The California TEAM study: Breath concentrations and personal exposures to 26 volatile compounds in air and drinking water of 188 residents of Los Angeles, Antioch, and Pittsburg, CA. *Atmospheric Environment* 22:2141-2163.

Wallace LA, Mitchell H, O'Connor GT, Neas L, Lippmann M, Kattan M, Koenig J, Stout JW, Vaughn BJ, Wallace D, Walter M, Adams K, Liu LJS. 2003. Particle concentrations in inner-city homes of children with asthma: The effect of smoking, cooking and outdoor pollution. *Environmental Health Perspectives* 111:1265-1272.

Wallace L, Ott W. 2011. Personal exposure to ultrafine particles. *Journal of Exposure Science & Environmental Epidemiology* 21(1):20-30.

Walsh M, Black A, Morgan A, Crawshaw GH. 1977. Sorption of SO_2 by typical indoor surfaces, including wool carpets, wallpaper and paint. *Atmospheric Environment* 11: 1107-1111.

Ward TJ, Palmer CP, Noonan CW. 2010. Fine particulate matter source apportionment following a large woodstove changeout program in Libby, Montana. *Journal of the Air & Waste Management Association* 60:688-693.

Wasson SJ, Guo ZS, McBrian JA, Beach LO. 2002. Lead in candle emissions. *Science of the Total Environment* 296:159-174.

Weschler CJ. 2000. Ozone in indoor environments: Concentration and chemistry. *Indoor Air* 10(4):269-288.

Weschler CJ. 2004. New directions: Ozone-initiated reaction products indoors may be more harmful than ozone itself. *Atmospheric Environment* 38:5715-5716.

Weschler CJ. 2006. Ozone's impact on public health: Contributions from indoor exposures to ozone and products of ozone-initiated chemistry. *Environmental Health Perspectives* 114:1489-1496.

Weschler CJ. 2009. Changes in indoor pollutants since the 1950s. *Atmospheric Environment* 43:153-169.

Weschler CJ, Nazaroff WW. 2008. Semivolatile organic compounds in indoor environments. *Atmospheric Environment* 42:9018-9040.

Weschler CJ, Nazaroff WW. 2010. SVOC partitioning between the gas phase and settled dust indoors. *Atmospheric Environment* 44:3609-3620.

WHO (World Health Organization). 1986. *WHO environmental health criteria 59: Principles for evaluating health risks from chemicals during infancy and early childhood: The need for a special approach*. Geneva: WHO Press.

WHO. 2007. *Indoor air pollution and lower respiratory tract infections in children*. Geneva: WHO Press.

Wilkinson P, Smith KR, Davies M, Adair H, Armstrong BG, Barrett M, Bruce N, Haines A, Hamilton I, Oreszczyn T, Ridley I, Tonne C, Chalabi Z. 2009. Public health benefits of strategies to reduce greenhouse-gas emissions: Household energy. *Lancet* 374:1917-1929.

Wilson KM, Klein JD, Blumkin AK, Gottlieb M, Winickoff JP. 2011. Tobacco-smoke exposure in children who live in multiunit housing. *Pediatrics* 127(1):85-92.

Wilson WE, Suh HH. 1997. Fine particles and coarse particles: Concentration relationships relevant to epidemiologic studies. *Journal of the Air & Waste Management Association* 47:1238-1249.

Wilson WE, Mage DT, Grant LD. 2000. Estimating separately personal exposure to ambient and nonambient particulate matter for epidemiology and risk assessment: Why and how. *Journal of the Air & Waste Management Association* 50:1167-1183.

Winickoff JP, Gottlieb M, Mello MM. 2010. Regulation of smoking in public housing. *New England Journal of Medicine* 362(24):2319-2325.

Wisthaler A, Weschler CJ. 2010. Reactions of ozone with human skin lipids: Sources of carbonyls, dicarbonyls, and hydroxycarbonyls in indoor air. *Proceedings of the National Academy of Sciences of the United States of America* 107:6568-6575.

Wolkoff P. 1995. Volatile organic compounds—Sources, measurements, emissions, and the impact on indoor air quality. *Indoor Air* 5(Suppl 3):1-73.

Wolkoff P, Clausen PA, Jensen B, Nielsen GD, Wilkins CK. 1997. Are we measuring the relevant indoor pollutants? *Indoor Air* 7:92-106.

Woodruff TJ, Axelrad DA, Caldwell J, Morello-Frosch R, Rosenbaum A. 1998. Public health implications of 1990 air toxics concentrations across the United States. *Environmental Health Perspectives* 106:245-251.

Wormuth M, Scheringer M, Vollenweider M, Hungerbühler K. 2006. What are the sources of exposure to eight frequently used phthalic acid esters in Europeans? *Risk Analysis* 26:803-824.

Yeh H-C, Cuddihy RG, Phalen RF, Chang I-Y. 1996. Comparisons of calculated respiratory tract deposition of particles based on the proposed NCRP model and the new ICRP66 model. *Aerosol Science and Technology* 25:134-140.

Ziska LH, Emche SD, Johnson EL, George K, Reed DR, Sicher RC. 2005. Alterations in the production and concentration of selected alkaloids as a function of rising atmospheric carbon dioxide and air temperature: Implications for ethno-pharmacology. *Global Change Biology* 11:1798-1807.

Ziska LH, Epstein PR, Schlesinger WH. 2009. Rising CO_2, climate change, and public health: Exploring the links to plant biology. *Environmental Health Perspectives* 117(2):155-158.

Ziska L, Knowlton K, Rogers C, Dalan D, Tierney N, Elder MA, Filley W, Shropshire J, Ford LB, Hedberg C, Fleetwood P, Hovanky KT, Kavanaugh T, Fulford G, Vrtis RF, Patz JA, Portnoy J, Coates F, Bielory L, Frenz D. 2011. Recent warming by latitude associated with increased length of ragweed pollen season in central North America. *Proceedings of the National Academy of Sciences of the United States of America* 108(10):4248-4251.

5

Dampness, Moisture, and Flooding

INTRODUCTION

This chapter addresses indoor environmental quality (IEQ) problems associated with moisture, condensation, and inundation and the possible effects of climate change on these problems. There is an extensive literature on the effects of indoor dampness on health, including an Institute of Medicine (IOM) report (IOM, 2004) that remains salient and is drawn on heavily in this chapter. The committee did not attempt to re-examine all the scientific evidence considered in the IOM report and other efforts—an undertaking beyond the scope of this study—but instead highlights their findings and other research relevant to the consideration of the health effects of alterations in IEQ induced by climate change.

The chapter's focus is on fungi[1] and bacteria—microbial agents that grow in the presence of water—and products of damaged building materials. They produce biologic and chemical emissions that can lead to irritant, allergic, other immunologic, or toxic responses. Other chapters address some issues relevant to occupants' exposures to those emissions. Ventilation, which is discussed in Chapter 8, has an effect on exposure: levels of indoor contaminants are higher in spaces that have lower air-exchange

[1] Fungi have eukaryotic cells as do animals and plants but are a separate kingdom. Most consist of masses of filaments, live off dead or decaying organic matter, and reproduce by spores. Visible fungal colonies found indoors are commonly called *mold* (mould), sometimes *mildew*. This report, following the convention of earlier IOM reports and much of the literature on indoor environments, uses the terms *fungus* and *mold* interchangeably to refer to the microorganisms.

rates. Some microbial agents also cause infections (this topic is addressed in Chapter 6).

CLIMATE CHANGE AND INDOOR DAMPNESS AND FLOODING

The effects of climate change on moisture indoors are driven by several factors, including extreme weather events, local changes in temperature and humidity, and the adaptations that occupants make and mitigation strategies that they use in response to changed environmental conditions.

The US Global Change Research Program notes that increases in air temperatures and increased frequency and intensity of heavy downpours have already been observed in the United States and that likely future changes "include more intense hurricanes with related increases in wind, rain, and storm surges" (USGCRP, 2009). Extreme weather conditions may lead to breakdowns in building envelopes followed by sudden infiltration of water into indoor spaces. Dampness problems and water intrusion create conditions favorable to the growth of fungi and bacteria and may cause building materials to decay or corrode and lead to off-gassing of chemicals. In areas where the climate is warm and humid for more months of the year, air conditioning will be used more often. Well-designed and properly operating heating, ventilation, and air conditioning (HVAC) systems can ameliorate humid conditions; poorly designed or maintained systems may introduce moisture and create condensation on indoor surfaces.[2] Mold-growth prevention and remediation may also introduce fungicides and other agents into the indoor environment, which can lead to adverse exposures of occupants.

Flooding as a result of extreme weather events can have profound health and economic effects. In 2010, there were 103 flood-related fatalities in the United States, a significantly higher number than the 10-year average of 71 measured between 2001 and 2010 (National Oceanic and Atmospheric Administration, 2011). In that same year, floods were part of six of the seven most costly insurance loss events in the United States; events that were responsible for $6.3 billion in losses (Swiss Re, 2011). Jonkman and colleagues (2009) estimate that two-thirds of the 771 known fatalities of Hurricane Katrina were the direct result of flooding and that additional fatalities were associated with flood-related circumstances including lack of access to potable water or medical services and exposure to extreme heat as a result of power outages.

Altered climatic conditions will not introduce new dampness problems into the indoor environment but may make existing problems more wide-

[2] This topic is addressed in Chapter 7.

spread and more severe and thus increase the urgency with which prevention and interventions must be pursued.

INDOOR DAMPNESS

Almost all buildings experience excessive moisture, leaks, or flooding at some point. Research regarding the sources and causes of indoor dampness was addressed in detail in a previous IOM report (2004), which described how and where buildings become wet; reviewed the signs of dampness, how dampness is measured, and what is known about its prevalence and characteristics, such as severity, location, and duration; discussed the risk factors for moisture problems; reviewed how dampness influences indoor microbial growth and chemical emissions; cataloged the various agents that may be present in damp environments; and addressed the influence of building materials on microbial growth and emissions. That effort's findings are briefly summarized below.

Dampness—a term used to describe a variety of moisture problems, including high relative humidity, condensation, water ponding, and other signs of excess moisture or microbial growth—is prevalent in residential housing. The prevalence and significance of dampness are less well understood in nonresidential structures, such as office buildings and schools, than in residential buildings.

There is no single cause of excessive indoor dampness, and the primary risk factors for it differ among climates, geographic areas, and building types. The prevalence of dampness problems appears to increase as buildings age and deteriorate, but some modern construction techniques and materials and the presence of air-conditioning can increase the risk of dampness problems. The prevalence and nature of these problems suggest that what is known about their causes and prevention is not consistently applied in building design, construction, maintenance, and use.

DAMPNESS AND HEALTH

Efforts to quantify the effects of indoor environmental factors on human health often rely on markers of dampness indoors to characterize risk. This approach reflects recognition that indoor moisture is associated with adverse health outcomes and that exposures to emissions from mold, bacteria, and damaged materials increase when indoor environments are chronically wet or damp.

There have been three large-scale reviews of the relationship between indoor dampness and human health in the past decade. In 2004, IOM issued *Damp Indoor Spaces and Health*. The World Health Organization (WHO) released *WHO Guidelines for Indoor Air Quality: Dampness and*

Mould in 2009, and researchers involved in the WHO effort updated and expanded that review in 2011 (Mendell et al., 2011).

The IOM report reviewed literature published up to late 2003 on a wide array of health effects. Among the major findings were that sufficient evidence existed for associating the presence of mold or other agents in damp buildings with nasal and throat symptoms, cough, wheeze, asthma exacerbation, and hypersensitivity pneumonitis in susceptible persons. The committee responsible for the IOM report concluded that limited or suggestive evidence existed for associating exposure to damp indoor environments with shortness of breath, asthma, and, in otherwise healthy children, lower respiratory disease. Tables 5-1 and 5-2 summarize the report's conclusions, and Box 5-1 summarizes the categories used to classify the strength of the evidence.

The WHO guidelines covered literature published up to July 2007 (WHO, 2009). Their authors took the same approach to evaluating and categorizing evidence for dampness as was used in the IOM report but ex-

TABLE 5-1 Summary of Findings Regarding the Association Between Health Outcomes and Exposure to Damp Indoor Environments[a]

Sufficient Evidence of a Causal Relationship	
(no outcomes met this definition)	
Sufficient Evidence of an Association	
Upper respiratory (nasal and throat) tract symptoms	Wheeze
Cough	Asthma symptoms in sensitized persons
Limited or Suggestive Evidence of an Association	
Dyspnea (shortness of breath)	Asthma development
Lower respiratory illness in otherwise healthy children	
Inadequate or Insufficient Evidence to Determine Whether an Association Exists	
Airflow obstruction (in otherwise healthy persons)	Skin symptoms
Mucous membrane irritation syndrome	Gastrointestinal tract problems
Chronic obstructive pulmonary disease	Fatigue
Inhalation fevers (nonoccupational exposures)	Neuropsychiatric symptoms
Lower respiratory illness in otherwise healthy adults	Cancer
Acute idiopathic pulmonary hemorrhage in infants	Reproductive effects
	Rheumatologic and other immune diseases

[a] The categories of evidence are summarized in Box 5-1 and explicated in *Damp Indoor Spaces and Health* (IOM, 2004).

TABLE 5-2 Summary of Findings Regarding the Association Between Health Outcomes and the Presence of Mold or Other Agents in Damp Indoor Environments[a]

Sufficient Evidence of a Causal Relationship	
(no outcomes met this definition)	
Sufficient Evidence of an Association	
Upper respiratory (nasal and throat) tract symptoms	Wheeze
Cough	Asthma symptoms in sensitized persons
Hypersensitivity pneumonitis in susceptible persons	
Limited or Suggestive Evidence of an Association	
Lower respiratory illness in otherwise healthy children	
Inadequate or Insufficient Evidence to Determine Whether an Association Exists	
Dyspnea (shortness of breath)	Skin symptoms
Asthma development	Gastrointestinal tract problems
Airflow obstruction (in otherwise healthy persons)	Fatigue
Mucous membrane irritation syndrome	Neuropsychiatric symptoms
Chronic obstructive pulmonary disease	Cancer
Inhalation fevers (nonoccupational exposures)	Reproductive effects
Lower respiratory illness in otherwise healthy adults	Rheumatologic and other immune diseases
Acute idiopathic pulmonary hemorrhage in infants	

[a] The categories of evidence are summarized in Box 5-1 and explicated in *Damp Indoor Spaces and Health* (IOM, 2004).

amined a larger set of health outcomes. Their analysis supported the IOM report findings that there was sufficient evidence to conclude that there is an association between indoor dampness-related agents[3] and asthma exacerbation, upper respiratory tract symptoms, cough, and wheeze. In addition, they determined that two outcomes not evaluated in the IOM report—current asthma and respiratory infections—and two outcomes that had been placed in the category of limited or suggestive evidence—asthma development and dyspnea (shortness of breath)—merited inclusion in the sufficient evidence category. Evidence regarding allergic rhinitis, bronchitis,

[3] Defined by the authors as "evidence of visible water damage, visible mold, mold odor, or similar related factors."

BOX 5-1
Summary of the Categories of Evidence Used in *Damp Indoor Spaces and Health* (IOM, 2004)

Sufficient Evidence of a Causal Relationship

Evidence is sufficient to conclude that a causal relationship exists between the agent and the outcome. That is, the evidence fulfills the criteria for "sufficient evidence of an association" and, in addition, satisfies the following criteria: strength of association, biologic gradient, consistency of association, biologic plausibility and coherence, and temporally correct association.

Sufficient Evidence of an Association

Evidence is sufficient to conclude that there is an association. That is, an association between the agent and the outcome has been observed in studies in which chance, bias, and confounding can be ruled out with reasonable confidence.

Limited or Suggestive Evidence of an Association

Evidence is suggestive of an association between the agent and the outcome but is limited because chance, bias, and confounding cannot be ruled out with confidence.

Inadequate or Insufficient Evidence to Determine Whether an Association Exists

The available studies are of insufficient quality, consistency, or statistical power to permit a conclusion regarding the presence of an association. Alternatively, no studies exist that examine the relationship.

and eczema—which had not been separately evaluated in the IOM report—was deemed limited or suggestive.

Mendell and colleagues carried the WHO review forward to late 2009. On the basis of their examination of previously available and newly published evidence, they raised bronchitis, allergic rhinitis, eczema, and ever-diagnosed asthma (that is, without regard to whether there was a current diagnosis of asthma) to the sufficient-evidence category. Epidemiologic research also yielded limited or suggestive evidence of an association between dampness-related agents and the "common cold" and "allergy/atopy."

The sections that follow provide some background on asthma, other respiratory ailments, and other conditions mediated by an immune response. They also highlight some of the recent literature on those health outcomes. Asthma is a prominent public-health concern because of rising rates and substantial effect on health, productivity, and health-care costs, but other immunologic conditions related to dampness are also problematic and may increase if sustained high levels of indoor moisture become more common (Mudarri and Fisk, 2007).

Asthma and Other Respiratory Ailments

In the United States, asthma from all causes increased in frequency and severity in the two decades from 1980 to 2000. From 1980 to 1996, the prevalence of asthma increased by 74%, and the incidence from 1.2 per 1,000 per year to 4.7 per 1,000 per year (Mannino et al., 2002). An assessment of annual asthma incidence in the total US population, using the National Health Interview Survey (NHIS) data for 1980–1996, estimated that 7.2–12.4% of those with prevalent asthma noted an onset in the preceding year (Rudd and Moorman, 2007). Data from the NHIS, the National Ambulatory Medical Care Survey, the National Hospital Ambulatory Medical Care Survey, the National Hospital Discharge Survey, and the National Vital Statistics System indicated that an estimated 8.2% of adults in the United States reported current asthma and that 4.2% of adults had at least one asthma attack in the previous year (Akinbami, 2011).

Efforts to estimate the burden of asthma that can be attributed to damp indoor spaces are limited by a lack of data on the prevalence of dampness indoors and by the absence of consistent occupational or environmental information on cases of asthma. Reviews estimate that one-fifth of current asthma in the United States is attributable to dampness in homes (Fisk et al., 2007) and that new-onset asthma or asthma-like symptoms may occur more frequently in people who are exposed to moisture or mold at home or at work (Sahakian et al., 2008). One study of office workers who occupied a water-damaged office building at a particular time documented an asthma incidence rate more than 7 times higher after occupancy than in the years before occupancy (1.9/1,000 person-years before building occupancy; 14.5/1,000 person-years after) (Cox-Ganser et al., 2005). Later analysis of that workforce with regard to exposure to mold, measured as culturable fungi and ergosterol concentrations in floor dust, demonstrated an excess risk of new-onset asthma at higher levels of exposure (Park et al., 2008).

Papers published since Mendell et al. (2011) completed their literature review in late 2009 have tended to support the conclusions drawn by them. A 2010 study of possible cases of occupational asthma in Finland determined that exposure to dampness and mold in the workplace was associated with new-onset adult asthma and aggravated the symptoms of asthmatics (Karvala et al., 2010). Nguyen and colleagues' analysis of the results of *The National Asthma Survey—New York State* found that there was a positive relationship between asthma symptoms, mold, and humidity in households that had at least one asthmatic adult or child (Nguyen et al., 2010).

A study in three urban cities in Korea established that students experienced higher levels of wheezing in classrooms that were damp, had visible mold growth, or had water damage (Kim et al., 2011). Sun et al. (2010) examined allergic symptoms, including wheezing, in students living in dormitories in Tianjin University, China, during the 2006–2007 school year. The students reported more moisture accumulation and moldy odors and higher levels of wheezing and rhinitis in summer than winter months. In contrast, Holme and colleagues' study of children in Sweden did not find an association between visible signs of dampness and spore concentration in indoor air or a relationship between spore concentrations and children's allergy and asthma symptoms (Holme et al., 2010).

Other Immunologic Conditions

Epidemiologic studies have shown that some immunologic outcomes in addition to asthma may be related to moisture incursion in buildings. Sarcoidosis is more frequent in occupants of water-damaged buildings (Cox-Ganser et al., 2005; Laney et al., 2009; Newman et al., 2004), including school buildings (Dangman et al., 2005). It is important to note that in each of the cited investigations of sarcoidosis, the researchers documented increases in asthma and asthma-like symptoms.

It is biologically plausible that exposure to bacteria (notably the endotoxin that is a cell-wall component of some bacteria) and fungi that are often present in damp indoor environments could trigger immune responses that lead to inflammation. Experimental studies have demonstrated that common microbial constituents of damp indoor environments can be potent inducers of inflammatory responses (Hirvonen et al., 2005). Researchers believe that granuloma formation in sarcoidosis is in response to an unidentified antigenic stimulus that induces a local Th1-cell–mediated immune response (DuBois et al., 2003). Chronic stimulation of macrophages causes the continuing release of inflammatory cytokines (IL-2, IL-12, IFN-c, and TNF-α), which leads to accumulation of Th1 cells at the site of inflammation. That immunologic cycling contributes to expansion of the granuloma structure (Richie, 2005).

Autoimmune diseases occur when a person mounts a specific immune response to self antigens that leads to tissue damage. Autoimmune diseases are often progressive and debilitating. The burden of autoimmune diseases in the United States is substantial: they affect an estimated 8% of the total population (Fairweather et al., 2008) and disproportionately affect females—more than three-fourths of cases of autoimmune diseases are in women (Dooley and Hogan, 2003; Gleicher and Barad, 2007; Jacobson et al., 1997). In a 2008 review, Fairweather et al. (2008) describe autoimmune diseases as the third-most common category of disease after cancer and cardiovascular disease in the United States. The role of environmental and occupational exposures is poorly defined (Gold et al., 2007), but exposure to external antigens may trigger and support an autoimmune inflammatory response (Münz et al., 2009), and joint symptoms and diseases have been associated with microbial exposures related to moisture damage (Luosujärvi et al., 2003).

SPECIFIC DAMPNESS-RELATED CONTAMINANTS

The principal indoor dampness-related agents that affect health are thought to be molds and bacteria that amplify in the presence of water and products of damaged building materials. As the *Damp Indoor Spaces and Health* report (IOM, 2004) notes, mold spores are regularly found in indoor air and on surfaces and materials; no indoor space is free of them. There are many species and genera, and those most typically found indoors vary in geographic area, climate, season, and other factors. The availability of moisture is the primary factor that controls mold growth indoors. Although much attention is focused on mold growth indoors, it is not the only dampness-related microbial agent. Mold growth is often accompanied by bacterial growth. Some research on fungi and bacteria focuses on specific components that may be responsible for particular health effects: hyphal (filament) fragments of fungi, protein allergens of microbial origin, structural components of fungal and bacterial cells, and such products as microbial volatile organic compounds (MVOCs) and mycotoxins. Release of those components depends on many physiologic and environmental factors. Dampness can also damage building materials and furnishings, causing or exacerbating the release of chemicals and other nonbiologic particles.

The following sections summarize information on those agents and some of the research on their affects on the health of building occupants.

Molds

Fungi exist as single cells (yeasts), filaments, fruiting bodies, and spores. They are composed of complex chemical compounds, including proteins, glycoproteins, glucans, and proteases. They produce cellular toxins in their competition for access to sources of nutrition in their environment, and their metabolic byproducts include volatile organic compounds (Bush and Portnoy, 2001; Storey et al., 2004).

Fungi are ubiquitous in nature and play a critical role in the natural decomposition of organic materials. Indoor spaces without moisture problems generally have air concentrations of mold that are the same as or lower than those outdoors, and the species are the same as those outdoors. Many fungal taxa in the indoor environment are similar to those recovered outdoors, but there are factors in the indoor environment (such as lack of fungicidal ultraviolet radiation from the sun, stable temperature, stable humidity, and shelter) that can allow some fungi to thrive better in the indoor environment.

Some structural components of mold can cause an immune response in a person who is exposed mainly through inhalation. Such responses are most commonly allergic and result in rhinitis or asthma. Other responses can lead to hypersensitivity pneumonitis (Ikeda et al., 2002; Lee et al., 2000; Patel et al., 2001; Seuri et al., 2000). The immunologic responses are complex, inasmuch as mold components include antigens and adjuvants that heighten the response to the antigens (Kheradmand et al., 2002; Reed, 2007).

Because the active agents that lead to those responses are macromolecules on the cell wall, fungal fragments are at least as likely to cause the reactions as are intact spores. Therefore, mold does not have to be living to have an immunologic effect on building occupants. In addition, various mold species share the macromolecules, so an allergy to one species results in an allergy to many (Green et al., 2005a, 2009; Schmechel, 2007).

MVOCs include alcohols, esters, aldehydes, and aromatic compounds. They cause the "musty" odor associated with moldy environments. They can cause irritation of mucous membranes (Horner and Miler, 2003), which can lead to irritation of the eyes, nose, throat, and respiratory tract. Irritation of the trigeminal nerve can lead to headache and fatigue. $(1\rightarrow3)$-β-D-glucans, components of cell-wall fragments, alter reactions to other agents (Rylander and Lin, 2000) and thus may add to the irritant properties described here.

Molds produce mycotoxins under some growth conditions (Jarvis, 2002). There are hundreds of those compounds (Etzel, 2002; Norred et al., 2001), and they include aflatoxins, fumonisins, ochratoxins, rubratoxins, and trichothecenes (Jarvis et al., 1995; Wannemacher and Wiener, 1997). Some have neurotoxic, cytotoxic, immunologic, reproductive, or carcinogenic properties. Although the compounds can exhibit severe toxicity in animals or humans when they are ingested or inhaled at high levels in, for example, agricultural settings, it is less clear whether they have an effect at the levels seen in occupied indoor spaces (IOM, 2004). Mycotoxins have been identified in building materials and settled dust in water-damaged buildings (Bloom et al., 2009). There is evidence that in these environments they contribute to inflammatory responses (Miller et al., 2010). Other potential effects are the subject of current investigation.

Not all dampness is the same for fungi. During Hurricane Katrina, wind-driven saltwater inundated many homes in Mississippi. The result was severe water damage, but the damage was different from that caused by the sustained floods in New Orleans from Lake Pontchartrain. After the water receded in Mississippi, the homes were dried, and mold growth was easily initiated on building materials and furnishings. New Orleans had homes that were essentially like sealed terrariums for several weeks at the end of summer 2005. A common scene in such buildings was a high-water mark on the drywall below which little mold growth was observed. There are two possible explanations for that difference. First, almost all molds need oxygen and cannot sporulate in liquid. That might explain why the previously submerged drywall had less mold growth; the time it took for the water to subside limited mold colonization. Second, flood waters contain the chemicals found in homes themselves (for example, bleach, pesticides, and other cleaning products), chemicals from the soil outside, and possibly other toxicants from nearby industrial or agricultural sources. Some of the chemicals can be fungus inhibitors or can be fungicidal. Perhaps some combination of the two reasons explains the pattern in homes that have endured long-term flooding.

The ramifications of long-term flooding could lead to differences in the types of fungi that can proliferate, but research is lacking. One study found that although *Cladosporium* spores and DNA were abundant and easily collected in air samples from homes, culturable colonies were not as common in heavily damaged homes in New Orleans (Chew et al. 2006). Given that *Cladosporium* is commonly recovered in home dust and air samples (Chew et al., 2003; Li and Kendrick, 1995; Wouters et al., 2000) and can easily compete with such species as *Aspergillus* and *Penicillium*, the lack of growing colonies was perplexing to the researchers.

The Centers for Disease Control and Prevention (CDC) has published detailed guidance on how to limit exposure to mold and how to identify and prevent mold-related health effects in the wake of hurricanes and floods (Brandt et al., 2006). It includes exposure-assessment instructions; remediation advice (including cleaning of HVAC systems); personal protective equipment recommendations for cleanup personnel; guidance on allergic, infectious, and toxic effects of exposure to mold and other dampness-related agents; adverse health-effects prevention strategies; and advice to public health-authorities. The authors recommend surveillance of community health after hurricanes and floods to identify unrecognized hazards and to gather information that will allow better responses in the future.

Few comprehensive epidemiologic studies have been conducted to assess respiratory effects of residents who lived in homes after major flooding. What is known is mainly from the Mississippi floods of 1993 and Hurricanes Katrina and Rita in 2005. Brown and colleagues (2006) estimate that in the New Orleans area alone the latter two events caused at least 110,000 homes to have high levels of mold and bacteria and at least 40,000 to be heavily contaminated.

Ross and colleagues assessed mold spores, lung function, and respiratory symptoms in 57 asthmatic residents of 44 homes in East Moline, Illinois, in April–October 1994 (Ross et al., 2000). The average mold-spore concentration was 2,190 spores/m^3. The researchers found that higher *Alternaria* concentrations were associated with missing sleep because of asthma (odds ratio [OR], 4.8; 95% Confidence Interval [CI], 1.6–14.6). In their second analysis of the data on the Mississippi floods, the researchers had a slightly different sample size; they assessed mold spores, lung function, and respiratory symptoms in 59 asthmatic residents of 46 homes in East Moline, Illinois, in April–October 1994 (Ross et al., 2002). Concentrations in this study averaged 5,692 spores/m^3. The researchers found that higher mold-spore concentrations were associated with an improved peak expiratory flow rate (PEFR) and respiratory symptom scores. They attribute the paradoxical results partly to self-reported diary cards for PEFR and symptoms.

Rabito et al. conducted two studies of mold exposure in post–Hurricane Katrina New Orleans. In the first, the study site was a school that reopened in January 2006, five months after the hurricane (Rabito et al., 2008). Respiratory health questionnaire and spirometric data were collected on children 7–14 years old, and air sampling for fungi in their homes was conducted at baseline and again after two months. The 75th percentile for mold concentration was 100 colony-forming units per cubic meter (cfu/m^3) and 70 cfu/m^3 at the two times. The concentrations were several orders of magnitude lower than those reported in unoccupied homes immediately after the hurricane (Chew et al., 2006). Nonetheless, there was an overall decrease in mold levels and respiratory symptoms over the study period, and indoor mold levels were low despite reported hurricane damage. Although many of the homes had sustained hurricane damage, the authors stressed that their results might not be generalizable to the residents of other homes who did not have the financial means to return to the city and to repair their homes or relocate to a nonflooded area.

In the other study by Rabito and colleagues, 529 patients in an allergy clinic were enrolled from December 1, 2005, to December 31, 2008. Mold exposure was assessed with a questionnaire, and mold allergy with a skin-prick test. Mold exposure (defined in terms of extent of home damage or duration of exposure) was not associated with mold allergy. The authors acknowledged that minorities and those without health insurance were underrepresented in the study, and this limited generalizability of the results.

Overall, the studies did not observe a statistically significant association between mold exposure and respiratory symptoms after flooding events. That result may be influenced by such factors as selection bias, lack of generalizability of the study populations, the healthy-resident effect (whereby healthy residents may be more able to conduct the necessary cleanup and renovation efforts), and difficulties in discerning associations between mold exposure and respiratory morbidity because of the presence of confounding factors (Barbeau et al., 2010).

Separately, Dales and colleagues (1991) used questionnaires to gather data on the health and home characteristics of more than 13,000 children

5–8 years old in 30 communities across Canada. Flooding, defined as "the appearance of flooding, water damage, or leaks in basement in last year," was associated with statistically significant ORs for parent-reported cough, wheeze, dyspnea, asthma, bronchitis, chest illness, upper respiratory symptoms, and eye irritation. The estimates were not adjusted for confounders, but the authors stated that analyses that adjusted for age, sex, race, parental education, presence of environmental tobacco smoke, presence of gas appliances, and hobbies that generate airborne contaminants yielded similar results.

Recovery activities after hurricane and floods also present risks. Cummings and colleagues (2008) found that people's respirator use while they were entering flooded areas and during cleanup and remediation decreased adverse exposures. They established that disposable-respirator use in water-damaged homes was associated with lower odds of exacerbation of moderate or severe upper respiratory symptoms (OR, 0.51; 95% CI, 0.24–1.09) and lower respiratory symptoms (OR, 0.33; 95% CI, 0.13–0.83).

Methods used to assess exposure to mold and mold components are a major area of research. For example, Ross et al. (2000) found that mold spores reflect a small fraction of the antigen load in a mold-contaminated space. Fungal fragments and conidia contribute allergens at concentrations orders of magnitude greater than mold spores (Green et al., 2005b, 2006). Airborne culturable fungi represent a yet smaller subset of the antigen load. One study assessing asthma morbidity in inner-city children with documented allergy to fungi and focusing on four genera of fungi found that elevated outdoor and indoor levels of culturable mold resulted in increased asthma morbidity (Pongracic et al., 2010). Further studies of the health impacts of post-flood events thus need to assess exposure using a number of methods, including qualitative characterization of contaminated surfaces and fungal fragments.

Bacteria

Bacteria also thrive in damp indoor environments and often coexist with mold. As noted earlier, they can cause inflammatory responses (Hirvonen et al., 2005). Endotoxin, a component of the cell wall of gram-negative bacteria, has been particularly well studied and has been shown either to have direct health effects or to augment the effects of other contaminants. Endotoxin in house dust has been associated with wheeze in infants (Keman et al., 1998; Park et al., 2001) and with greater severity of asthma in adults who are sensitive to dust mites (Michel, 1996; Michel et al., 1991). In the workplace, endotoxin has been found at high levels in association with hypersensitivity pneumonitis (Rose et al., 1998), and levels in floor dust have been shown to be associated with lower and upper respiratory symptoms, fever and chills, and headache in a large office building and to interact with fungi in the floor dust and lead to higher rates of lower respiratory symptoms in occupants who have increased fungal and endotoxin levels (Park et al., 2006).

Emissions from Damaged Building Materials and Furnishings

Water damage can lead to decay of building materials and furnishings. Polyvinyl chloride (PVC) floor coatings release phthalates when exposed to water, and phthalates in house dust have been associated with allergic symptoms, eczema, and asthma in children (Bornehag et al., 2004; Jaakkola and Knight, 2008). Increased rates of asthma and allergy symptoms have been associated with damp PVC floor coatings (Bornehag et al., 2005; Tuomainen et al., 2004). Understanding of the complex chemical interactions that occur indoors is growing. Research has shown that the indoor chemistry of surfaces (vinyl tile, wall board, and carpet) and the gas phase reactions that can occur when surfaces are disturbed can result in the rapid formation of potential irritants (Forester and Wells, 2009; Ham and Wells, 2008; Harrison and Ham, 2009; Wells et al., 2008). Indoor surfaces can also be important reservoirs of reactant chemicals—such as cleaning agents, pesticides, and paints—that can undergo hydrolysis reactions because of moisture. Characterization of those exposures and their associated health effects is a subject of active research (Anderson et al., 2007, 2010).

SUMMARY COMMENTS

Dampness problems in buildings are pervasive, and strategies for avoiding them well established, although not necessarily widely implemented. There are several sources of guidance on design and retrofit strategies. Lstiburek (2004, 2005a,b, 2006), for example, has produced a series of books that offer design and construction advice specific to various housing types and climatic conditions found in the United States, including advice on avoiding water intrusion and excessive indoor dampness. Operational advice—in particular, proper operation of HVAC systems—for avoiding damp conditions indoors is also available (ASHRAE, 2009).

The 2004 IOM report *Damp Indoor Spaces and Health* summarizes dampness and mold remediation guidelines issued by the New York City Department of Health and Mental Hygiene (NYCDOH 1993, 2000), Health Canada (1995), the American Conference of Governmental Industrial Hygienists (ACGIH, 1999), the US Environmental Protection Agency (EPA, 2001), and the American Industrial Hygiene Association (AIHA, 2001). CDC also offers advice based on the Occupational Safety and Health Administration, EPA, and New York City 2005 revised guidance (Brandt et al., 2006).

The summary advice of those authors and organizations is straightforward. Quoting from *Damp Indoor Spaces and Health* (IOM, 2004):

- Homes and other buildings should be designed, operated, and maintained to prevent water intrusion and excessive moisture accumulation when possible. When water intrusion or moisture accumulation is discovered, the sources should be identified and eliminated as soon as practicable to reduce the possibility of problematic microbial growth and building material degradation.
- When microbial contamination is found, it should be eliminated by means that limit the possibility of recurrence and limit exposure of occupants and persons conducting the remediation.

Operationalizing the advice, however, is difficult. The 2004 IOM report committee concluded that "the prevalence and nature of dampness problems suggest that what is known about their causes and prevention is not consistently applied in building design, construction, maintenance, and use." Buildings are a complex combination of foundation, structure elements, and interior components, including insulation, plumbing, HVAC, and ancillary systems. Changes in one may affect the function of others in ways that are difficult to anticipate.

Climate change may complicate dampness prevention planning and responses. Buildings are—at least ideally—designed to operate in a particular set of outdoor environmental conditions. Local building codes are predicated on those conditions, specifying resistance against projected weather extremes. Building-insurance interests base their premium calculations (and their economic viability) on assumptions regarding the ability of the structures that they underwrite to survive such extremes. If climatic conditions in a particular area change—for example, if there are more severe or more frequent episodes of intense precipitation—buildings constructed under existing codes and designed to operate under previously existing conditions may fail to perform as designed under the new conditions. That suggests that careful consideration must be given to revising building codes and practices to anticipate future climatic conditions and to taking a coordinated approach to addressing risks.

CONCLUSIONS

On the basis of its review of the papers, reports, and other information presented in this chapter, the committee has reached the following conclusions regarding the health effects of alterations in IEQ due to dampness and flooding:

- Studies reviewed in the 2004 IOM report *Damp Indoor Spaces and Health* and confirmed by research indicate that
  - o Excessive indoor dampness is a determinant of the presence or source strength of several potentially problematic exposures. Damp indoor environments favor house-dust mites and the growth of mold and other microbial agents, standing water supports cockroach and rodent infestations, and excessive moisture may initiate or enhance chemical emissions from building materials and furnishings.
  - o Damp indoor environments are associated with the initiation or exacerbation of a number of respiratory ailments.
- Extreme weather and flooding events that penetrate buildings—which may become more frequent or severe in the future—increase the number of people at risk for health conditions related to standing water, wet building materials, and sustained high indoor humidity.
- Dampness problems in buildings can be difficult to anticipate. The information needed to minimize the risk of their occurrence or their severity is available but is not being consistently applied.
- Current buildings and building design, construction, operation, and maintenance practices may not be appropriate for managing indoor dampness or flooding problems due to outdoor environmental conditions that could result from climate change. New, flexible approaches that anticipate potential problems and take measures to prevent them or minimize their adverse consequences are needed.

REFERENCES

ACGIH (American Conference of Governmental Industrial Hygienists). 1999. *Bioaerosols—Assessment and control*. Cincinnati, Ohio.

AIHA (American Industrial Hygiene Association). 1996. *Field guide for the determination of biological contaminants in environmental samples*. Fairfax, VA: AIHA Press.

AIHA. 2001. *Report of the Microbial Task Force*. Fairfax, VA: AIHA Press.

Akinbami L. 2011. Asthma prevalence, health care use, and mortality: United States, 2005-2009. *National Health Statistics Reports 32*.

ASHRAE (American Society of Heating, Refrigerating and Air-Conditioning Engineers). 2009. *Indoor air quality guide: The best practices for design, construction and commissioning*. Atlanta, GA: ASHRAE.

Anderson SE, Wells JR, Fedorowicz A, Butterworth LF, Meade BJ, Munson AE. 2007. Evaluation of the contact and respiratory sensitization potential of volatile organic compounds generated by simulated indoor air chemistry. *Toxicological Sciences* 97:355-363.

Anderson SE, Jackson LG, Franko J, Wells JR. 2010. Evaluation of dicarbonyls generated in a simulated indoor air environment using an in vitro exposure system. *Toxicological Sciences* 115:453-461.

Barbeau DN, Grimsley LF, White LE, El-Dahr JM, Lichtyeld M. 2010. Mold exposure and health effects following hurricanes Katrina and Rita. *Annual Review of Public Health* 31:165-178.

Bloom E, Grimsley LF, Pehrson C, Lewis J, Larsson L. 2009. Molds and mycotoxins in dust from water-damaged homes in New Orleans after Hurricane Katrina. *Indoor Air* 19(2):153-158.

Bornehag CG, Sundell J, Sigsgaard T. 2004. Dampness in buildings and health (DBH): Report from an ongoing epidemiological investigation on the association between indoor environmental factors and health effects among children in Sweden. *Indoor Air* 14(Suppl 7):59-66.

Bornehag CG, Sundell J, Hagerhed-Engman L, Sigsgaard T, Janson S, Aberg N, DBH Study Group. 2005. "Dampness" at home and its association with airway, nose, and skin symptoms among 10,851 preschool children in Sweden: A cross-sectional study. *Indoor Air* 15(Suppl 10):48-55.

Brandt M, Brown C, Burkhart J, Burton N, Cox-Ganser J, Damon S, Falk H, Fridkin S, Garbe P, McGeehin M, Morgan J, Page E, Rao C, Redd S, Sinks T, Trout D, Wallingford K, Warnock D, Weissman D. 2006. Mold prevention strategies and possible health effects in the aftermath of hurricanes and major floods. *MMWR Recommendations and Reports* 55(RR-8):1-27.

Brown C., Riggs M, Rao C, Cumming K. 2006. *Health concerns associated with mold in water-damaged homes after Hurricanes Katrina and Rita—New Orleans area, Louisiana, October 2005*. 38th Joint Meeting—Panel on Wind and Seismic Effects, 15-20 May 2006. http://www.pwri.go.jp/eng/ujnr/joint/38/paper/38-56brown.pdf (accessed February 13, 2011).

Bush RK, Portnoy JM. 2001. The role and abatement of fungal allergens in allergic diseases. *Journal of Allergy and Clinical Immunology* 107(3):430-442.

CDC (Centers for Disease Control and Prevention). 2006. *Behavioral Risk Factor Surveillance System Survey data*. Atlanta, GA: Centers for Disease Control and Prevention.

Chew GL, Rogers C, Burge HA, Muilenberg ML, Gold DR. 2003. Dustborne and airborne fungal propagules represent a different spectrum of fungi with differing relations to home characteristics. *Allergy* 58(1):13-20.

Chew GL, Wilson J, Rabito FA, Grimsley F, Iqbal S, Reponen T, Muilenberg ML, Thorne PS, Dearborn DG, Morely RL. 2006. Mold and endotoxin levels in the aftermath of Hurricane Katrina: A pilot project of homes in New Orleans undergoing renovation. *Environmental Health Perspectives* 114(12):1883-1889.

Cox-Ganser JM, White SK, Jones R, Hilsbos K, Storey E, Enright PL, Rao CY, Kreiss K. 2005. Respiratory morbidity in office workers in a water-damaged building. *Environmental Health Perspectives* 113(4):485-490.

Cummings, KJ, Cox-Ganzer JM, Riggs MA, Edwards N, Hobbs GR, Kreiss K. 2008. Health effects of exposure to water-damaged New Orleans homes six months after Hurricanes Katrina and Rita. *American Journal of Public Health* 98(5):869-875.

Dales R, Zwanenburg H, Burnett R, Franklin C. 1991. Respiratory health effects of home dampness and molds among Canadian children. *American Journal of Epidemiology* 134(2):196-203.

Dangman KH, Bracker AL, Storey E. 2005. Work-related asthma in teachers in Connecticut: Association with chronic water damage and fungal growth in schools. *Connecticut Medicine* 69(1):9-17.

Dooley MA, Hogan SL. 2003. Environmental epidemiology and risk factors for autoimmune disease. *Current Opinion in Rheumatology* 15(2):99-103.

Du Bois RM, Goh N, McGrath D, Cullinan P. 2003. Is there a role for microorganisms in the pathogenesis of sarcoidosis? *Journal of International Medicine* 253(1):4-17.

EPA (US Environmental Protection Agency). 2001. *Mold remediation in schools and commercial buildings*. Washington, DC: EPA Office of Air and Radiation.

Etzel RA. 2002. Mycotoxins. *Journal of the American Medical Association* 287(4):425-427.

Fairweather DL, Frisancho-Kiss S, Rose NR. 2008. Sex differences in autoimmune disease from a pathological perspective. *The American Journal of Pathology* 173(3):600-609.

Fisk WJ, Lei-Gomez Q, Mendell MJ. 2007. Meta-analyses of the associations of respiratory health effects with dampness and mold in homes. *Indoor Air* 17:284-296.

Forester CD, Wells JR. 2009. Yields of carbonyl products from gas-phase reactions of fragrance compounds with OH radical and ozone. *Environmental Science and Technology* 43(10):3561-3568.

Gleicher N, Barad DH. 2007. Gender as risk factor for autoimmune diseases. *Journal of Autoimmunity* 28(1):1-6.

Gold LS, Ward MH, Dosemeci M, De Roos AJ. 2007. Systemic autoimmune disease mortality and occupational exposures. *Arthritis & Rheumatism* 56(10):3189-3201.

Green BJ, Schmechel D, Sercombe JK, Tovey ER. 2005a. Enumeration and detection of aerosolized *Aspergillus fumigatus* and *Penicillium chrysogenum* conidia and hyphae using a novel double immunostaining technique. *Journal of Immunological Methods* 307(1-2):127-134.

Green BJ, Sercombe JK, Tovey ER. 2005b. Fungal fragments and undocumented conidia function as new aeroallergen sources. *Journal of Allergy and Clinical Immunology* 115(5):1043-1048.

Green BJ, Tovey ER, Sercombe JK, Blachere FM, Beezhold DH, Schmechel D. 2006. Airborne fungal fragments and allergenicity. *Medical Mycology* 44(Suppl 1):S245-S255.

Green BJ, Tovey ER, Beezhold DH, Perzanowski MS, Acosta LM, Divjan AI, Chew GL. 2009. Surveillance of fungal allergic sensitization using the fluorescent halogen immunoassay. *Journal of Medical Mycology* 19(4):253-261.

Ham JE, Wells JR. 2008. Surface chemistry reactions of α-terpineol [(R)-2-(4-methyl-3-cyclohexenyl)isopropanol] with ozone and air on a glass and a vinyl tile. *Indoor Air* 18:394-407.

Harrison JC, Ham JE. 2009. β–Ionone reactions with the nitrate radical: Rate constant and gas-phase products. *International Journal of Chemical Kinetics* 41(10):629-641.

Health Canada. 1995. *Fungal contamination in buildings: A guide to recognition and management*. http://individual.utoronto.ca/jscott/fpwgmaqpb001.pdf (accessed February 14, 2011).

Hirvonen MR, Huttunen K, Roponen M. 2005. Bacterial strains from moldy buildings are highly potent inducers of inflammatory and cytotoxic effects. *Indoor Air* 15(Suppl 9):65-70.

Holme J, Hägerhed-Engman L, Mattsson J, Sundell J, Bornehag CG. 2010. Culturable mold in indoor air and its association with moisture-related problems and asthma and allergy among Swedish children. *Indoor Air* 20(4):329-340.

Horner WE, Miller JD. 2003. Microbial volatile organic compounds with emphasis on those arising from filamentous fungal contaminants from buildings. *ASHRAE Transactions: Research* 4621 (RP-1072).

Ikeda T, Kuroda M, Ueshima K. 2002. A case of hypersensitivity pneumonitis caused by *Gyrodontium versicolor*. *Nihon Kokyuki Gakkai Zasshi* 40(5):387-391.

IOM (Institute of Medicine). 2004. *Damp indoor spaces and health*. Washington, DC: The National Academies Press.

Jaakkola JJ, Knight DL. 2008. The role of exposure to phthalates from polyvinyl chloride products in the development of asthma and allergies: A systematic review and meta-analysis. *Environmental Health Perspectives* 116(7):145-163.

Jacobson DL, Gange SJ, Rose NR, Graham NM. 1997. Epidemiology and estimated population burden of selected autoimmune diseases in the United States. *Clinical Immunology and Immunopathology* 84(3):223-243.

Jarvis BB. 2002. Chemistry and toxicology of molds isolated from water-damaged buildings. *Advances in Experimental Medicine and Biology* 504:43-52.

Jarvis BB, Salemme J, Morals A. 1995. *Stachybotrys* toxins. 1. *Natural Toxins* 3(1):10-16.

Jonkman SN, Maaskant B, Boyd E, Levitan ML. 2009. Loss of life caused by the flooding of New Orleans after Hurricane Katrina: Analysis of the relationship between flood characteristics and mortality. *Risk Analysis* 29(5):676-698.

Karvala K, Toskala E, Luukkonen R, Lappalainen S, Uitti J, Nordman H. 2010. New-onset adult asthma in relation to damp and moldy workplaces. *International Archives of Occupational and Environmental Health* 83(8):855-865.

Keman S, Jetten M, Douwes J, Borm PJA. 1998. Longitudinal changes in inflammatory markers in nasal lavage of cotton workers: Relation to endotoxin exposure and lung function changes. *International Archives of Occupational and Environmental Health* 71(2):131-137.

Kheradmand F, Rishi K, Corry DB. 2002. Environmental contributions to the allergic asthma epidemic. *Environmental Health Perspectives* 110(Suppl 4):553-556.

Kim JL, Elfman L, Wislander G, Ferm M, Torén K, Norbäck D. 2011. Respiratory health among Korean pupils in relation to home, school and outdoor environment. *Journal of Korean Medical Science* 26(2):166-173.

Laney AS, Cragin LA, Blevins LZ, Sumner AD, Cox-Ganser JM, Kreiss K, Moffatt SG, Lohff CJ. 2009. Sarcoidosis, asthma, and asthma-like symptoms among occupants of a historically water-damaged office building. *Indoor Air* 19(1):83-90.

Lee SK, Kim SS, Nahm DH, Park HS, Oh YJ, Park KJ, Kim SO, Kim SJ. 2000. Hypersensitivity pneumonitis caused by *Fusarium napiforme* in a home environment. *Allergy* 55(12):1190-1193.

Li DW, Kendrick B. 1995. A year-round study on functional relationships of airborne fungi with meterological factors. *International Journal of Biometeorology* 39(2):74-80.

Lstiburek J. 2004. *Builder's guide to hot-dry/mixed-dry climates*. Westford, MA: Building Science Corporation.

Lstiburek J. 2005a. *Builder's guide to hot/humid climates*. Westford, MA: Building Science Corporation.

Lstiburek J. 2005b. *Builder's guide to mixed-humid climates*. Westford, MA: Building Science Corporation.

Lstiburek J. 2006. *Builder's guide to cold climates*. Westford, MA: Building Science Corporation.

Luosujärvi RA, Husman TM, Seuri M, Pietikäinen MA, Pollari P, Pelkonen J, Hujakka HT, Kaipiainen-Seppänen OA, Aho K. 2003. Joint symptoms and diseases associated with moisture damage in a health center. *Clinical Rheumatology* 22(6):381-385.

Mannino DM, Homa DM, Akinbami LJ, Moorman JE, Gwynn C, Redd SC. 2002. Surveillance for asthma—United States, 1980-1999. *MMWR Surveillance Summary* 51(1):1-13.

Mendell MJ, Mirer AG, Cheung K, Tong M, Douwes J. 2011. Respiratory and allergic health effects of dampness, mold, and dampness-related agents: A review of the epidemiologic evidence. *Environmental Health Perspectives* 119(6): doi:10.1289/ehp.1002410.

Michel O, Ginanni R, Duchateau J, Vertongen F, le Bon B, Sergysels R. 1991. Domestic endotoxin exposure and clinical severity of asthma. *Clinical & Experimental Allergy* 21(4):441-448.

Michel O. 1996. Endotoxin and asthma. *Revue Française d'Allergologie et d'Immunologie Clinique* 36(8):942-945.

Miller JD, Sun M, Gilyan A, Roy J, Rand TG. 2010. Inflammation-associated gene transcription and expression in mouse lungs induced by low molecular weight compounds from fungi from the built environment. *Chemico-Biological Interactions* 183(1):113-124.

Mudarri D, Fisk JW. 2007. Public health and economic impact of dampness and mold. *Indoor Air* 17(3):226-235.

Münz C, Lünemann JD, Teague Getts M, Miller SD. 2009. Antiviral immune responses: Triggers of or triggered by autoimmunity? *Nature Reviews Immunology* 9:246-258.

National Oceanic and Atmospheric Administration (NOAA). 2011. *Natural hazard statistics.* http://www.weather.gov/om/hazstats.shtml (accessed May 2, 2011).

Newman LS, Rose CS, Bresnitz EA, Rossman MD, Barnard J, Frederick M, Terrin ML, Weinberger SE, Moller DR, McLennan G, Hunninghake G, DePalo L, Baughman RP, Iannuzzi MC, Judson MA, Knatterud GL, Thompson BW, Teirstein AS, Yeager H Jr., Johns CJ, Rabin DL, Rybicki BA, Cherniack R, ACCESS Research Group. 2004. A case control etiologic study of sarcoidosis: Environmental and occupational risk factors. *American Journal of Respiratory and Critical Care Medicine* 170(12):1324-1330.

Nguyen T, Lurie M, Gomez M, Reddy A, Pandya K, Medvesky M. 2010. The National Asthma Survey–New York State: Association of the home environment with current asthma status. *Public Health Reports* 125(6):877-887.

Norred WP, Riley RT, Meredith FI, Poling SM, Plattner ID. 2001. Instability of *N*-acetylated fumonisin B1 (FA1) and the impact on inhibition of ceramide synthase in rat liver slices. *Food and Chemical Toxicology* 39(11):1071-1078.

NYCDOH (New York City Department of Health and Mental Hygiene). 1993. *Assessment and remediation of* Stachybotrys atra *in indoor environments.* New York City: Department of Health.

NYCDOH. 2000. *Guidelines on assessment and remediation of fungi in indoor environments.* New York City: Department of Health.

Park JH, Spiegelman DL, Gold DR, Burge HA, Milton DK. 2001. Predictors of airborne endotoxin in the home. *Environmental Health Perspectives* 109(8):859-864.

Park JH, Cox-Ganzer J, Rao C, Kreiss K. 2006. Fungal and endotoxin measurements in dust associated with respiratory symptoms in a water-damaged office building. *Indoor Air* 16(3):192-203.

Park JH, Cox-Ganzer JM, Kreiss K, White SK, Rao CY. 2008. Hydrophilic fungi and ergosterol associated with respiratory illness in a water-damaged building. *Environmental Health Perspectives* 116(1):45-50.

Patel AM, Ryu JH, Reed CE. 2001. Hypersensitivity pneumonitis: Current concepts and future questions. *Journal of Allergy and Clinical Immunology* 108(5):661-670.

Pongracic JA, O'Connor GT, Muilenberg ML, Vaughn B, Gold DR, Kattan M, Morgan WJ, Gruchalla RS, Smartt E, Mitchell HE. 2010. Differential effects of outdoor versus indoor fungal spores on asthma morbidity in inner-city children. *Journal of Allergy and Clinical Immunology* 125(3):593-599.

Rabito FA, Iqbal S, Kiernan MP, Holt E, Chew GL. 2008. Children's respiratory health and mold levels in New Orleans after Katrina: A preliminary look. *Journal of Allergy and Clinical Immunology* 121(3):622-625.

Reed CE. 2007. Inflammatory effect of environmental proteases on airway mucosa. *Current Allergy and Asthma Reports* 7(5):368-374.

Richie RC. 2005. Sarcoidosis: A review. *Journal of Insurance Medicine* 37(4):283-294.

Rose CS, Martyny JW, Newman LS, Milton DK, King TE Jr., Beebe JL, McCammon JB, Hoffman RE, Kreiss K. 1998. "Lifeguard Lung": Endemic granulomatous pneumonitis in an indoor swimming pool. *American Journal of Public Health* 88(12):1795-1800.

Ross MA, Curtis L, Scheff PA, Hryhorczuk DO, Ramakrishnan V, Wadden RA, Persky VW. 2000. Association of asthma symptoms and severity with indoor bioaerosols. *Allergy* 55(8):705-711.

Ross MA, Persky VW, Scheff PA, Chung J, Curtis L, Ramakrishnan V, Wadden RA, Hryhorczuk DO. 2002. Effect of ozone and aeroallergens on the respiratory health of asthmatics. *Archives of Environmental Health* 57:568-578.

Rudd RA, Moorman JE. 2007. Asthma incidence: Data from the National Health Interview Survey, 1980-1996. *The Journal of Asthma* 44(1):65-70.

Rylander R, Lin RH. 2000. (1→3)-β-D-glucan—Relationship to indoor air-related symptoms, allergy and asthma. *Toxicology* 152(1-3):47-52.

Sahakian NM, White SK, Park JH, Cox-Ganzer JM, Kreiss K. 2008. Identification of mold and dampness-associated respiratory morbidity in 2 schools: Comparison of questionnaire survey responses to national data. *Journal of School Health* 78(1):32-37.

Schmechel D, Lindsley WG, Chen TB, Blachere FM, Green BJ, Brundage RA, Beezhold DH. 2007. A two-stage personal cyclone sampler for the collection of fungal aerosols and direct ELISA and PCR sample analysis. *Journal of Allergy and Clinical Immunology* 119(1/Suppl 1):S188.

Seuri M, Husman K, Kinnunen H, Reiman M, Kreus R, Kuronen P, Lehtomäki K, Paananen M. 2000. An outbreak of respiratory diseases among workers at a water-damaged building—A case report. *Indoor Air* 10:138-145.

Storey E, Dangman K, Schenck P, DeBernardo R, Yang C, Bracker A, Hodgson M. 2004. *The recognition and management of health effects related to mold exposure and moisture indoors*. Storrs, CT: Center for Indoor Environments and Health at University of Connecticut Health Center.

Sun Y, Zhang Y, Bao L, Fan Z, Sundell J. 2010. Ventilation and dampness in dorms and their associations with allergy among college students in China: A case-control study. *Indoor Air* doi: 10.1111/j.1600-0668.2010.00699.x.

Swiss Re. 2011. *Natural catastrophes and man-made disasters in 2010: A year of devastating and costly events*. Zurich: Swiss Reinsurance Company Ltd. Economic and Research Counseling.

Tuomainen A, Seuri M, Sieppi A. 2004. Indoor air quality and health problems associated with damp floor coverings. *International Archives of Occupational and Environmental Health* 77(3):222-226.

USGCRP (US Global Change Research Program). 2009. *Global climate change impacts in the United States*. New York: Cambridge University Press.

Wannemacher RW Jr., Wiener SL. 1997. Trichothecene mycotoxins. In *medical aspects of chemical and biological warfare (Textbook of military medicine. Part 1, Warfare, weaponry, and the casualty, V. 3.)*, edited by Sidell FR, Takafuji ET, Franz DR. Washington, DC: Office of the Surgeon General, Borden Institute, Walter Reed Army Medical Center.

Wells JR, Morrison GC, Coleman BK. 2008. Kinetics and reaction products of ozone and surface-bound squalene. *Journal of ASTM International* 5(7):JAI101629.

WHO (World Health Organization). 2009. *WHO guidelines for indoor air quality: Dampness and mould*. Copenhagen: WHO Regional Office for Europe.

Wouters IM, Douwes J, Doekes G, Thorne PS, Brunekreef B, Heederik DJ. 2000. Increased levels of markers of microbial exposure in homes with indoor storage of organic household waste. *Applied and Environmental Microbiology* 66(2):627-631.

6

Infectious Agents and Pests

Many pathogens and allergens are profoundly affected by environmental conditions. Their survival may be directly influenced by temperature, humidity, or moisture, or their availability may depend on the distribution, abundance, or behavior of their hosts or vectors. A changing climate will thus affect human exposure to these agents.

This chapter addresses indoor environmental quality concerns associated with the infectious agents and other pests that research suggests may be influenced by climate-change–induced alterations in the indoor environment. The chapter also touches on exposure to chemicals used to control pest infestations. Exposures that are directly related to dampness are the subject of Chapter 5.

Two earlier National Academies reports have addressed issues relevant to the material discussed in this chapter. The 2001 National Research Council report *Under the Weather: Climate, Ecosystems, and Infectious Disease* (NRC, 2001) and the 2008 Institute of Medicine workshop summary *Global Climate Change and Extreme Weather Events* (IOM, 2008a) take on the larger question of the linkages among climate, ecosystems, and infectious disease. A white paper commissioned by the US Environmental Protection Agency (EPA) in conjunction with the present effort discusses the potential effects of climate change on microbial air quality in the built environment (Morey, 2010).

INFECTIOUS AGENTS

Infectious diseases have been major drivers of evolution and of human evolution in particular. The vast majority of infections are acquired from the environment or transmitted from humans or other animals. Therefore, factors that affect the physical environment, how we build in it, and how we share it with other humans are critical determinants of the infections to which we are exposed and how we perpetuate the exposures. Seasonal variation—a complex summing of multiple influences ranging from sunlight to moisture to wind speed and varying by region—has also been recognized as a critical influence on infectious-disease epidemiology dating back to Hippocrates (Naumova, 2006). Thus, climate change in general and indoor-air exposure in particular are major elements in the spread or interruption of infectious diseases in humans. Despite the extensive knowledge base on the effects of climate change on environmental growth of microorganisms and their vectors and hence infections, however, data on the effects of climate change on indoor air and infectious diseases are incomplete.

This section briefly reviews some of the most pertinent model systems that highlight elements of the knowledge in direct effects of climate on infectious disease. It explores them by category of infection, inasmuch as each kingdom (for example, bacteria, fungi, and viruses) has distinct features and is involved in different processes and exposures. One critical factor is that air and moisture, and therefore water, are inextricably linked. Most microorganisms are exquisitely sensitive to moisture, either requiring it or avoiding it. Therefore, the study of indoor air is closely linked to the state of indoor water, its aerosols, and the magnitude of humidity. Furthermore, pipes and other water-delivery systems are prone to development of biofilms, thin, removal-resistant layers of metabolically inaccessible bacteria that are constantly available for release into water and indoor air through taps, showers, humidifiers, and the like.

Respiratory Viruses

Experience dating back thousands of years has taught that infectious diseases can be affected by seasonal changes; this suggests that environment plays a critical role in the modulation of disease load, spread, and susceptibility. Obvious and recurring examples are provided by the respiratory viruses, most notably influenza viruses, respiratory syncytial virus (RSV), and the rhinoviruses. Mechanisms of spread are varied and include aerosol, fomite,[1] and direct contact. Direct contact, such as hand-to-hand transfer, is the most easily modified and is a major contributor to the spread of respira-

[1] Fomites are inanimate objects or substances—a door knob, for example—that function to transfer infectious organisms from one individual to another.

tory viruses. Fomite spread is affected by ambient humidity, which can in turn be affected by indoor air.

Influenza Viruses

Influenza viruses continue to account for substantial annual morbidity and mortality interspersed with periods of increased activity. The 2009–2010 H1N1 influenza epidemic is estimated to have involved around 61 million infections, 274,000 hospitalizations, and more than 12,000 deaths in the United States (CDC, 2010b).

Although there has been prolonged controversy over the environmental correlates of influenza epidemic spread, it appears that absolute humidity—the amount of water vapor in a given volume of air—is a critical determinant (Shaman and Kohn, 2009; Shaman et al., 2010a,b). In contrast, relative humidity—the amount of water vapor in a given volume of air at a given temperature expressed as the percentage of the maximum possible for that temperature—is well regulated in the indoor environment and appears not to be as important a determinant of influenza transmission and spread. However, studies by Myatt et al. (2010) show that increased absolute humidity and relative humidity, achieved by the use of indoor air humidification, can lead to substantial reductions in viable influenza virus.[2] Overall, the effects of humidity on influenza virus outbreaks and peak epidemic periods are greater in temperate than in tropical environments. In some tropical and subtropical settings, relative humidity has been more closely associated with influenza epidemics (Tang et al., 2010a,b). Because periods of high relative humidity corresponded to periods of increased indoor time and air conditioning, the population-based correlations are confounded. However, because indoor air conditioning affects indoor temperature and humidity, these require more investigation to determine whether the critical aspects of influenza spread are determined by the indoor or outdoor environmental conditions. The different results in temperate and tropical zones may reflect differences in viral and human biology in those regions. However, comparative studies for tropical and subtropical regions for respiratory transmission have not been completed in the United States.

Respiratory Syncytial Virus

RSV is the greatest cause of bronchiolitis and pneumonia in infants worldwide and causes up to about 125,000 hospitalizations in US in-

[2] As discussed later in this chapter, though, increased humidity may create a more hospitable environment for mold growth and accelerate the degradation and subsequent off-gassing of building materials and furnishings.

fants each year. In the US elderly population, it accounts for an estimated 177,000 hospitalizations and 14,000 deaths (CDC, 2010). RSV appears to contribute to invasive pneumococcal disease more than influenza viruses do (Murdoch and Jennings, 2009; Talbot et al., 2005; Watson et al., 2006).

Like influenza virus activity, RSV activity is highest in temperate climates during fall and winter months and into spring. However, there can be variability in the time of onset and duration, at least in more subtropical regions (CDC, 2010a). Unlike influenza virus, RSV is stabilized by higher humidity, and transmission in some studies correlates with relative humidity, lower temperature, and increased cloud cover (Meerhoff et al., 2009). Whether the mechanisms of these factors are due to direct effects on the virus or to indirect effects in driving people indoors into crowded environments is an open question. In some settings, such as Indonesia, RSV activity correlated strongly with rainfall and temperature (Omer et al., 2008). However, the apparently differing epidemiology in temperate and tropical climates remains incompletely explained (Welliver, 2009). In Spain, RSV admissions of infants with severe disease were strongly associated with lower temperature and lower absolute humidity (Lapeña et al., 2005).

Rhinovirus

Human rhinovirus (HRV) is a common and relatively mild pathogen, but one that by its very ubiquity and frequency has a major impact on human health, especially in the setting of pre-existing airway disease like asthma. Adults may have up to four bouts per year, typically in the fall through spring, accounting for up to 62 million cases in the United States annually (Sloan et al., 2011). In addition, because HRV is highly transmissible, settings that favor human-to-human and fomite transmission tend to result in relatively high rates of HRV during certain times of the year. Less research has been conducted on HRV than on influenza and RSV, in part because these latter organisms' morbidity and mortality are much higher and their etiologies somewhat less complex.

Human rhinoviruses are comprised of three main groups—A, B, and C—which replicate in the epithelial cells of the upper and lower respiratory tracts, leading to cough, wheeze, and rhinorrhea (Dulek and Peebles, 2011). Allergic triggers act along with HRV to fuel the exacerbation of asthma. Extensive work has shown that HRV is one of the most prevalent cofactors in asthma exacerbations, making their role in overall medical care critical to understand and interrupt.

A few studies address the determinants of HRV transmission and prevalence in indoor environments. Myatt et al. (2004) showed that the amount of HRV recovered from building air handling filters varied with the amount of outside air entrained, suggesting that HRV transmission might be influenced by the number of air exchanges in the work environment. Singleton and colleagues (2010) found that HRV was recovered from 44% of Alaskan native children hospitalized with a respiratory infection, but this rate was quite close to that in control children who were not hospitalized. Tovey and Rawlinson (2011) note that the rates of asthma rise precipitously two to three weeks after the start of school, indicating that some new exposure in the classroom is responsible. The authors hypothesize that these factors include HRV as well as numerous other costimulators of asthma such as endotoxin, proteins, and allergens. du Prel and colleagues (2009) found that HRV rates are associated with higher humidity levels, which might become more common as a result of climate change.

Gram-Negative Bacteria

The gram-negative bacteria present special issues in climate-associated infectious-disease epidemiology. They are not dependent on human-to-human spread, are not dependent on human inhabitation for survival, and have the ability to form biofilms—slippery, poorly penetrable slimes that cover the inside of water conduits. Given their close ties to the environment and their access to humans through water consumption, aerosol generation, heating, and cooling, the epidemiology of gram-negative rod infections is a window into infectious diseases in the setting of climate change.

Legionella

From its initial recognition as a cause of human respiratory disease, *Legionella* infection has been closely tied to water-droplet exposure in hotels and hospitals (Stout and Yu, 1997). However, the modes of transmission clearly can involve both aerosol spread (by water misters in grocery stores, for example) and aspiration. Spread from potting soil has also been well documented (de Jong and Zucs, 2010).

Regardless of the exposures or the modes of transmission, it is clear that legionellae are relatively common in some water supplies and has seasonal variation. In a case-crossover study in the greater Philadelphia area, Fisman and colleagues identified summertime occurrence of reported *Legionella* pneumonia to correlate with rainfall and increased relative humidity in the preceding week or so, rather than temperature (Fisman et al., 2005). Whether that reflects increased recruitment of legionellae into the water supply through rainfall, increased survival in higher humidity, indoor transmission, or outdoor transmission remains to be concretely determined. However, it is clear that in many instances, such as in hospitals, *Legionella* transmission is presaged by high levels of bacterial or bacterial DNA recovery from ambient water sources, such as faucets (Feazel et al., 2009). This dynamic reservoir of organisms probably serves as the source of aerosol generation, the source of bacteria that can be aspired by predisposed hosts, or both. Thus, indoor water clearly influences *Legionella* transmission and is itself influenced by regional environmental factors.

Pseudomonas aeruginosa

Stapleton et al. (2007) studied the incidence and causes of keratitis in contact-lens wearers in Australia. They found that *Pseudomonas aeruginosa* accounted for a plurality of cases and that it varied with higher mean minimum temperature but not humidity. Conducting their study in a country with well-characterized tropical and more temperate zones, they found that although *P. aeruginosa* was most common in the tropical regions, gram-positive organisms, such as *Staphylococcus aureus,* predominated in more temperate regions (Stapleton et al., 2007). Perencevich et al. (2008) studied the effects of seasonal temperature on nosocomial infection rates at the University of Maryland Medical Center. On review of almost 218,594 cases and 26,624 unique cultures, they found that rates of some gram-negative bacillary infections, including *P. aeruginosa* infections, were higher during warmer months and that rates of *P. aeruginosa* infection increased in relation to temperature rise. Gram-negative organisms that showed similar seasonal variation included *Acinetobacter baumanii*, *Enterobacter cloacae*, and *Escherichia coli*. Rates of gram-positive bacteria, such as *S. aureus* and *Enterococcus* spp., were not increased over the same periods and did not show similar relationships to temperature. Those infections occurred in hospitals, so they are reflections of effects of indoor environment, but they presumably reflect some changes in the outdoor environment as well. That other nosocomial pathogens, such as *S. aureus*, did not vary in the same pattern excludes simple effects of climate on human practices and suggests a more intrinsic effect of climate on gram-negative nosocomial pathogens.

As mentioned above, Perencevich et al. (2008) showed that gram-negative nosocomial infections increased with increasing temperature in Baltimore. In a national survey, McDonald et al. (1999) also found seasonal variation in *Acinetobacter baumanii* nosocomial infections but not in *P. aeruginosa* infections. They also noted marked differences in regional rates of *A. baumanii* infections, with higher rates in the eastern than western parts of the United States.

Mycobacteria

The Mycobacteriaceae are typically environmentally hardy gram-positive rods that include the high-grade primate pathogen *Mycobacterium tuberculosis,* the more numerous environmental or nontuberculous mycobacteria, and *M. leprae*, the agent of leprosy. Some of these organisms have emerged as agents of lung infection in patients who have underlying lung diseases that lead to impaired clearance of respiratory secretions. Those diseases are best exemplified by cystic fibrosis, a genetic disease in which impairment and dysfunction of the airway-lining cilia lead to the airway-widening condition known as bronchiectasis. Bronchiectasis is a common feature of the other syndromes in which nontuberculous mycobacterial infections occur, including primary ciliary dyskinesia, alpha-1 antitrypsin deficiency, and hyper-IgE recurrent-infection syndrome (Zoumot and Wilson, 2010).

The role of environmental exposure, including exposure to the indoor environment, in nontuberculous mycobacterial infection has recently received intense interest. The nontuberculous mycobacteria live in temperate and tropical waters and soils throughout the world. Unlike *M. tuberculosis* and *M. leprae*, which depend almost exclusively on human-to-human spread for their propagation, the nontuberculous mycobacteria are environmental opportunists that live in biofilms and can survive otherwise hostile environments because of their waxy cell walls (Falkinham, 2010). Feazel et al. (2009) showed recovery of *M. avium* complex genetic signatures from biofilms collected from inside showerheads in homes. Other organisms were also detected, including legionellae. Falkinham et al. (2008) reported a case of pulmonary infection with a particular species of *M. avium* complex that was recovered from the home water supply; this suggested spread from the household water to the patient. That potential mechanism of spread has been expanded on by Chan and Iseman (2010). The occurrence of pulmonary nontuberculous mycobacterial infection is highest in cystic fibrosis patients who have the mildest forms of disease, especially in women (Rodman et al., 2005).

Fomites

Increasing relative humidity and temperature outdoors will probably lead to increased indoor dampness and dampness-related health effects. As is the case with many infectious-disease vectors, the effects of temperature and relative humidity may increase or decrease the survival of viruses and bacteria and facilitate the persistence of infectious fomites (Boone and Gerba, 2007). Increases in environmental temperature decrease the survival of many viruses. For example, the H5N1 avian influenza virus persisted on duck feathers and on surfaces for long times but only at lower temperatures (Wood et al., 2010; Yamamoto et al., 2010). The combination of a stable indoor environment and increased dampness may actually decrease the transmission of some respiratory viruses and increase the survival of other pathogens on fomites, such as the ones that harbor bacteria and mold

(Boone and Gerba, 2007; Gubler et al., 2001). Increased dampness indoors, possibly exacerbated by building deterioration, may exacerbate or increase the risk of developing select respiratory diseases caused by mold and bacterial exposure (IOM, 2004; WHO, 2007).

Fungi

Fungi pose a special set of problems because they are ubiquitous, grow easily in the environment, and cause human diseases. However, the language surrounding fungal interactions with humans is fraught with imprecision, which leads to confusion. In addition, there are several distinct types of fungi, including yeasts, molds, and dimorphic yeasts (fungi that live as yeasts under one set of circumstances but can act like molds in other circumstances) (Holland and Vinh, 2009). The distinctions are important because the dimorphic yeasts are able to live both in the environment and in humans and cause some degree of invasive disease even in healthy humans. Examples include *Histoplasma capsulatum*, *Coccidioides immitis*, *Blastomyces dermatitidis*, *Sporothrix schenkii,* and *Paracoccidioides brasiliensis.* Those agents are relatively regional in their distribution and are therefore often referred to as endemic fungi. In healthy hosts, they can cause usually self-limited respiratory illnesses, such as valley fever due to *C. immitis*. They are organisms that live in the upper layer of soil outdoors and are rarely associated with indoor exposures and have rarely associated with indoor exposures to date. However, a white paper commissioned by EPA (Morey, 2010) suggests a mechanism by which this could change. It indicates that the upper layer of soil is prone to disturbance by dust storms, which may become more common in geographic areas that experience drought because of shifts in climatic conditions. This may in turn lead to greater indoor penetration of pathogenic fungi contained in soil and to higher indoor exposures in the absence of enhanced HVAC filtration or settled dust removal.

Invasive fungal infections are quite rare in humans and occur almost exclusively in the setting of immunocompromise, either inborn, such as some primary immunodeficiencies, or acquired, such as that acquired through transplantation or chemotherapy. However, with the advent of more drugs that affect immunity, such as tumor-necrosis factor alpha–(TNF-α)-blocking agents used for rheumatoid arthritis and inflammatory bowel disease, the number of people at risk for the development of fungal infection is increasing (Tsiodras et al., 2008). In susceptible persons *Aspergillus fumigatus*, a thermotolerant filamentous mold, can cause invasive disease that is usually spread by inhalation. Pneumonias that occur in the setting of immunocompromise carry high morbidity and mortality.

In the nonimmunocompromised host, the most important fungal disease in the respiratory tract is allergic bronchopulmonary aspergillosis

(ABPA), a syndrome of allergic response to fungi that is most common in those who are atopic, those who have cystic fibrosis, and those who have asthma (Patterson and Strek, 2010). Allergic fungal sinusitis is similar in that eosinophil-rich secretions become dense and involved with fungi without tissue invasion; in this case, the organisms involved include the dematiaceous (dark-walled) molds *Bipolaris spicifera* and *Curvularia lunata* or *Aspergillus fumigatus*, *A. niger*, and *A. flavus* (Schubert, 2009). The syndromes of chronic fungal rhinosinusitis are regionally concentrated around the South and Southwest of the United States. These allergic respiratory syndromes straddle the lines between infection, colonization, and allergy.

Synthesis

Climate change has many effects on infectious diseases, some malign and some ameliorative. How we adapt the indoor environment to the continuing changes in the outdoor environment will be critical determinants of how we affect the occurrence and spread of infectious diseases. In particular, effects on moisture, temperature, and the organisms trafficked into our homes, places of work, hospitals, and schools in water will determine the rates of viral, bacterial, mycobacterial, fungal, and allergic diseases.

PESTS

Indoor environments contain a number of unwelcome insects, other arthropods, and invasive animals. All of these are at some level sensitive to environmental conditions, but some are more susceptible to the conditions associated with climate change. This section summarizes the available literature on the characteristics of these pests; the health effects of exposure to the allergens and microbial agents that they produce, host, or carry; and how climate-change–induced alterations in the indoor environment—including changes in occupant behavior—may affect adverse exposures associated with them.

House Dust Mites

House dust mites are microscopic arthropods that are ubiquitous in indoor environments. They are among the most important sources of allergens in house dust and of allergic disease in the United States (IOM, 2000).

Exposure

Voorhorst and colleagues were the first to show that dust mites of the genus *Dermatophagoides* were the source of "house dust" allergens

(Voorhorst et al., 1969). *Dermatophagoides pteronyssinus* (de Boer and Kuller, 1997; van Strien et al., 1994; Voorhorst et al., 1969) and *D. farinae* (Antens et al., 2006) are commonly recovered in home settings. *D. farinae* is the hardier of the two (Arlian, 1975; Arlian and Veselica, 1981). An intervention study showed that the major allergen from *D. pteronyssinus* may have been decreased by an extremely dry (and cold) winter during the study period rather than by the home interventions themselves (Brunekreef et al., 2005; Gehring et al., 2005).

Dust mite viability is highly influenced by environmental conditions. There may be some inferences that as the climate warms, dust mites will thrive (Ayres et al., 2009). That is not entirely true. As noted in Chapter 2, although some regions of the country will experience warmer climates, they will not necessarily experience higher humidity. The critical factor for dust mites is water activity (A_w), which is relative humidity at a surface. Dust mites do not have lungs that can condition the air; rather, they conduct transpiration through their exoskeletons. A decrease in ambient relative humidity (which is paralleled by a drop in A_w) can affect dust mites not only in laboratory settings (Arlian, 1975; Arlian and Veselica, 1981) but in the home (Arlian et al., 2001; Cabrera et al., 1995; Harving et al., 1994) and at a community level (Acosta et al., 2008; Chew et al., 1999).

New York and Boston are coastal cities, but many of their homes can be dry in winter, and this factor eradicates the dust mite population. Studies indicate that increased indoor temperature in those communities has not been accompanied by an observed increase in the dust mite population; rather, dust mites decreased (Acosta et al., 2008; Chew et al., 1999). The homes where overheating was measured in these studies were multifamily apartment buildings whose residents had little control over their heating. The heating was turned on (building wide) early in fall and turned off late in spring. Figures 6-1 and 6-2 illustrate how overheated apartments compared with single-family homes whose residents had more control over their heating.

A change in climate could also affect the ecologic niches of some types of dust mites in such a way that the geographic patterns of endemic dust mites could change. As discussed earlier, some dust mites are more sensitive to humidity than others. The Dutch intervention study described earlier (Brunekreef et al., 2005) showed not only that dust mite levels decreased in this coastal country but that the profile of dust mite taxa had changed. Although it was not highlighted in the study, careful examination of one of the figures shows that between the beginning of the study (1996) and eight years later, *Der f* 1 (the major allergen from *D. farinae*) apparently became the most highly concentrated allergen in house dust (Antens et al., 2006). Even if humidity does not change substantially, warmer climate patterns are predicted, and this (in the absence of any adaptation measures, such as

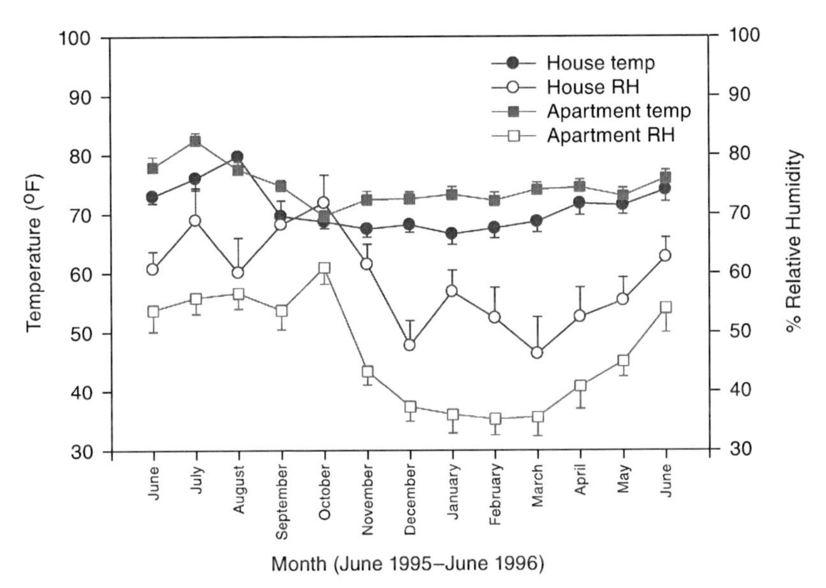

FIGURE 6-1 Variations in indoor temperature and relative humidity as functions of housing type and time of year in a sample of urban residences. (Derived from data presented in Chew et al., 1999.)

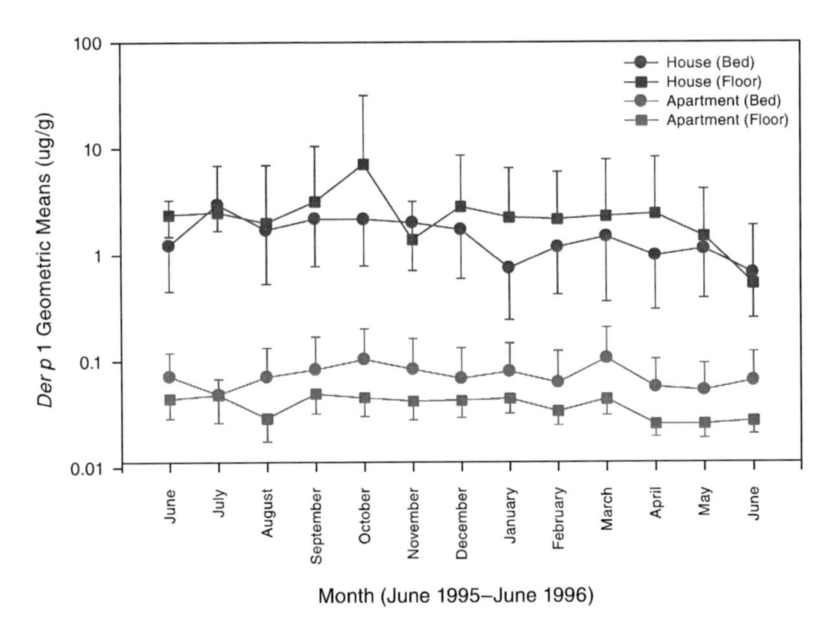

FIGURE 6-2 Variations in *Der p* 1 allergen levels as a function of housing type and location and time of year in a sample of urban residences. (Derived from data presented in Chew et al., 1999.)

increased use of air conditioning and dehumidifiers) could lead to a shift in the types of dust mites that are found in northern latitudes. The long-term outlook would be for increased numbers of dust mites and a wider variety of dust mite species in northern climates. However, that is speculative and does not consider adaptation measures.

Adaptation

More air-conditioning and dehumidifier use in summer In Ohio, Arilan et al. (2001) found that among three groups—those with only opening windows for cooling purposes, those with only air conditioning, and those with air conditioning and a high-efficiency dehumidifier—dust mite allergen concentrations in the homes with both air conditioning and dehumidifiers were less than 10% of concentrations in the other homes. Given that the monthly outdoor temperature in the Ohio summer months averaged 20.5–25.9°C (69–79°F), it is not clear how often the air conditioning was used in the air conditioning–only group. For that reason alone, the dehumidification (with only air conditioning) might not have been as much as could be expected in a scenario in which Ohio summers would be longer and warmer. Nonetheless, in a study that covered a wider geographic area of the United States, air conditioning was independently associated with lower dust mite allergen levels (Lintner and Brame, 1993). Furthermore, Swedish researchers found that mechanical exhaust and adequate supply ventilation in energy-efficient housing can decrease mite allergen concentrations substantially (Sundell et al., 1995). Air-conditioning use in the United States has steadily increased over the past several years, and it is likely that this adaptation measure will affect dust mite populations.

More humidifier use in summer Humidifier use is an adaptation strategy for those in arid environments. A humidifier would need to run quite often to keep the humidity high enough to sustain a dust mite population, and there is some evidence that they are in some circumstances. Prasad et al. (2009) showed that evaporative coolers can increase indoor relative humidity to a point where the prevalence of patients who have asthma or allergic rhinitis with sensitization to dust mites increases. The researchers posit that it is an increase in dust mite allergen in the environment that is tied with the increase in sensitization, but they collected no allergen measurements. Nonetheless, it remains an interesting research question that warrants further investigation.

Less heating in winter (because winters are milder) If winters become milder, some residents might use their heat less in the winter, and this could give rise to more dust mites and possibly a shift in the types of dust mites that can thrive in a particular climate. For example, if the northeast United States becomes warmer and more humid, such mites as *Blomia tropicalis*—which is found in Puerto Rico—might extend their range northward in the United States (Acosta, 2008; Montealegre et al., 1997).

Epidemiology

Current dust mite sensitization patterns in the United States Most of what is known about current dust mite sensitization patterns in the United States is from the National Health and Nutrition Examination Survey (NHANES) population-based sample of the general population and from childhood-asthma studies (Table 6-1). The NHANES data did not focus on dust mite sensitization more widely than the regional level. However, the National Cooperative Inner-City Asthma Study (NCICAS), Childhood Asthma Management Program (CAMP), and Inner-City Asthma Study (ICAS) show that, among children who have asthma, dust mite sensitization varied by city.

How allergen avoidance could affect sensitization patterns Early-life environmental exposures are key in the development of allergic sensitization (Illi et al., 2006; Lau et al., 2000). It is controversial which factors are most important, such as early-life pet ownership and early-life endotoxin exposure (Holt and Thomas, 2005), but it is clear that those with a genetic predisposition to inhalant allergies tend to become sensitized to what is in their environment (Chew et al., 2008; Eldeirawi et al., 2005; Huss et al., 2001; Ingram et al., 1995; Montealegre et al., 2004; Phipatanakul et al., 2000a,b). The CAMP showed that of children who had asthma in Albuquerque, Baltimore, Boston, Denver, St. Louis, San Diego, Seattle, and Toronto, those in Denver and Albuquerque had the lowest concentrations of dust mites in their homes and the lowest prevalence of dust mite sensitization (33.1% and 21.5%, respectively) (Huss et al., 2001). If climate change leads to longer heat waves, areas with climates similar to that of Albuquerque could expand, and residents in those areas would experience not only decreased allergen exposure (which would lead to fewer symptoms) but changes in the pattern of their allergic sensitization.

How asthma incidence patterns could be affected Generally speaking—that is, without regard to the sensitizing agent(s)—asthma mortality and morbidity rates exhibit geographic, regional, and seasonal differences in the United States. Rates are elevated in urban areas when compared to rural areas (Carr et al., 1992; Grant et al., 1999; Weiss and Wagner, 1990), with New York City, Chicago, and Phoenix having consistently high mortality and hospitalization rates (Carr et al., 1992; Lang and Polansky, 1994;

TABLE 6-1 Prevalence of Dust Mite Sensitization

Study	Type of Study	Population of Interest	Sample Size	Fraction Sensitized
NHANES II and III (Arbes et al., 2005)	Population-based survey of general population	Children and adults (6–59 years old)	17,738	4.2% (*D. farinae*)
NHANES III (Stevenson et al., 2001)	Population-based survey of general population	Children (6–16 years old)	4,164	25% (*D. farinae*)
CAMP (Huss et al., 2001)	Multisite asthma study	Children (5–12 years old)	1,041	48.5% (*D. pteronyssinus* or *D. farinae*)
NCICAS (Kattan et al., 1997)	Multisite asthma study	Children (4–9 years old)	1,286	24% (*D. farinae*), 31% (*D. pteronyssinus*)
ICAS (Gruchalla et al., 2005)	Multisite asthma study	Children (5–11 years old)	937	61.8% (*D. pteronyssinus* or *D. farinae*)
Tucson Children's Respiratory Study (Stern et al., 2004)	Birth cohort (Tucson)	Children (11 years old)	626	14.9% (*D. farinae*)
Epidemiology of Home Allergens and Asthma Study (TePas et al., 2006)	Birth cohort (Boston)	Children (7 years old)	131	34% (*D. pteronyssinus* or *D. farinae*)

NOTE: CAMP = Childhood Asthma Management Program, ICAS = Inner-City Asthma Study, NCICAS = National Cooperative Inner-City Asthma Study, NHANES =National Health and Nutrition Examination Survey.

Marder et al., 1992). Asthma mortality rates were consistently higher in the West and Midwest US Census regions than in the Northeast or South over the years 1980–1998 (Mannino, 2002). Akinbami (2006) notes that asthma prevalence rates among children 0–17 years of age are generally higher in the Northeast region than elsewhere but cautions:

> While it is tempting to attribute prevalence patterns to climate or air quality, many factors affect prevalence and may also vary by region. Some examples include the likelihood that symptomatic children are diagnosed accurately with asthma and population composition. For example, the Puerto Rican population, in which asthma prevalence is highest, tends to be concentrated in the Northeast region of the country.

Exacerbations follow the seasonal patterns exhibited by asthma co-morbidities, including rhinovirus and other respiratory viral infections, and triggers like pollen and mold (Johnston and Sears, 2006). Peak exacerbation occurs during the fall, although its magnitude varies depending on the age of the subject (younger asthmatics are more sensitive to seasonal changes).

Studies suggest that climate change will take place over a long period, and "allergen avoidance" might not be as extreme as that in some study interventions. Boner et al. (2002) found that dust mite–sensitized children who had asthma and were moved temporarily to the Italian Alps had reduced morbidity, and this was attributed to the absence of dust mites. Morgan et al. (2004) found that a targeted allergen-avoidance strategy for children reduced their asthma symptoms and emergency-department visits over a period of two years. What is more likely is that with the changing pattern of dust mite–endemic areas and the change in allergic-sensitization patterns, people who have allergic asthma will mount immune responses to elements of their environment, such as cockroaches, cats, and mice (Gruchalla et al., 2005). In contrast, if the northern states experience milder winters and an increase in humidity, dust mites might become the dominant allergen and surpass cockroaches and pets as the allergen most associated with increased asthma morbidity (Chew et al., 2009).

Other Pests

Little information is available on the potential effects of climate change on indoor exposure to other pests. Research has noted that the presence of increased mammalian pests in the indoor environment can spread disease and exacerbate allergies (IOM, 2008b), and increased outdoor temperatures are thought to have brought rodents indoors and led, for example, to disease from exposure to hanta virus in mouse droppings (Gubler et al., 2001). It is plausible that climate change will engender other indoor exposures to pests, but research on the question is lacking. It should be noted, however, that data collected as part of the National Survey of Lead and Allergens in Housing (Cohn et al., 2004) and National Cooperative Inner-City Asthma Study (Phipatanakul et al., 2000a,b; Platts-Mills et al., 2007; Pongracic et al., 2008) do not indicate that there are appreciable geographic differences in mouse allergen levels measured indoors. It is thus unclear whether climate variations may have an effect on rodent infestations.

Pest Controls

Several factors may cause pesticide exposure to increase under conditions of climate change. Higher temperatures may lead to increased numbers of structural, agricultural, and forest insect pests (Boxall et al., 2009; Quarles, 2007). Water and storm damage may expand opportunities for pests to invade the indoor environment (Brennan, 2010). And damage caused by flooding and an increase in available water or moisture could create environments that are more hospitable to pests and increase the capacity of buildings to support infestations. All those circumstances are likely to lead to greater residential and agricultural use of pesticides to control increasing populations of insects, rodents, and other disease vectors and thus to a greater risk of exposures of populations.

Vulnerable Populations

Some communities appear to be at greater risk for indoor pesticide exposure than others. In the United States, urban communities are particularly at risk. In New York state, a study found that the heaviest application of pesticides occurred not in agricultural counties but in the New York City boroughs of Manhattan and Brooklyn (Thier et al., 1998). Some 93% of residents in public housing in New York City reported applying pesticides in their homes, and more than half said that they did so once a week (Surgan et al., 2002). Bradman et al. (2005) found that rodent infestation in homes increased in the presence of peeling paint, water damage, and high residential density, and the use of pesticides is common in communities that have adverse housing conditions.

Extensive work in the past decade has demonstrated the presence of pesticides in urban house dust and addressed the risk of exposure to these chemicals in vulnerable groups such as pregnant women and children. Such factors as multifamily dwellings, leaky buildings that allow pests to come into the indoor environment, and home ownership influence the likelihood of pesticide use (Julien et al., 2008). Although regular application of chemical pesticides, some of which include banned or restricted products, may be used to combat severe infestation, the methods are often deemed ineffective.

Once introduced into the home environment, pesticide residues may persist for years beyond the time of application, as demonstrated by evidence of banned compounds, such as DDT, in residential dust samples (Julien et al., 2008; Stout et al., 2009).

Integrated pest management (IPM) is a reasonable adaptation to increasing populations of pests, but many families lack the education or resources to implement the changes that will be needed. Residents of multifamily urban dwellings where pesticides are commonly used may have little control over the pesticides used in their buildings.

Changes in Vector Distribution

Changes in patterns of infestation in the outdoor environment may affect indoor air quality. Milder and shorter winters are expected to increase the geographic distribution of pests, such as mosquitoes and insects that attack agricultural crops (Quarles, 2007). Outbreaks of West Nile virus, carried by mosquitoes, in 2001–2005 correlated with increasing temperature and rainfall, and this leads to the expectation that such outbreaks will increase with climate change. Others have drawn attention to how outbreaks of disease, such as dengue fever and possibly malaria, could result from climate change (Girman, 2010; Hales et al., 2002; Randolph and Rogers, 2000).

The change in vector distribution and increasing threats to the agriculture industry are expected to result in increased use of agricultural chemicals (Boxall et al., 2009). As outdoor pesticide applications increase in an attempt to control increasing distribution of pests, indoor levels of these contaminants could rise. Pesticides applied in the outdoor environment do not remain outdoors but can find their way indoors through air exchange or can be brought in on clothing, skin, and especially shoes. People who live close to agricultural operations that increase their use of chemicals to control insect infestation may be at particularly high risk (Ward et al., 2006). In orchard-producing areas of Washington and Oregon, pesticide levels in house dust have been associated with distance from agricultural fields (Lu et al., 2000; McCauley et al., 2001), and pesticide metabolites in urine increased with proximity to the fields and during the pesticide-application season (Lu et al., 2000). Levels of pesticides metabolites have been higher in the urine of agricultural children than in the urine of children who reside in nonagricultural communities (Lambert et al., 2005; Lu et al., 2000).

In countries where malaria is endemic, the residential ban on applications of DDT is being lifted. In 2006, the World Health Organization and the US Agency for International Development endorsed indoor DDT spraying to control malaria (WHO, 2009). The increasing distribution of pests associated with climate change will result in increased measures to control outbreaks and potentially increase the risk of exposure to pesticides and of associated health effects, particularly in vulnerable populations.

Health Concerns

Biologic monitoring in the United States indicates widespread exposure to organochlorine and organophosphate pesticides in the general population (National Exposure Research Library, 2005). Prospective cohort studies of mothers and newborns have documented considerable pesticide exposure during pregnancy in urban populations, with insecticides detected in air samples and in blood samples from women and newborns at delivery (Berkowitz et al., 2003; Whyatt et al., 2002, 2003). Those findings raised concern about potential health effects of residential exposure to pesticides. Many pesticides are developed to degrade quickly in outdoor environments but sequester in indoor environments in the absence of sunlight and rain. Pesticides can pass through the blood–brain barrier and penetrate the placenta. In addition, young children may receive greater exposure than adults because they eat, drink, and breathe more per unit of body weight (NRC, 1993). Children are also particularly vulnerable because they play in the dirt and on the floor (Fenske et al., 1990; Zwiener and Ginsburg, 1988).

Numerous animal studies have demonstrated that in utero or early exposure to organophosphate pesticides affects neurodevelopment (Eskenazi et al., 1999). Fetuses and young children may be more susceptible to neurotoxic effects of pesticides and have lower than adult levels of enzymes that are needed to detoxify organophosphate pesticides (Furlong et al., 2006). An emerging literature provides evidence of neurobehavioural consequences of relatively small exposure to organochlorine and organophosphate pesticides in infants and children (Eskenazi et al., 2008). Recent studies have found that the levels of organophosphate pesticides in dwellings may be great enough to cause neurodevelopmental effects (Eskenazi et al., 2007; Perera et al., 2003).

The increasing use of DDT to control malaria poses important questions about potential health effects. DDT is extremely persistent in the environment, so the potential association with indoor air quality is of concern. In a recent review of health effects associated with DDT exposure, a consensus group concluded that indoor residual spraying can result in substantial exposure to DDT and that DDT exposure may pose a risk to human populations (Eskenazi et al., 2009). There is a growing body of evidence that exposure to DDT and its breakdown product DDE may be associated with adverse health outcomes, such as breast cancer, diabetes, decreased semen quality, spontaneous abortion, and impaired neurodevelopment in children. However, few studies have measured body burdens of both DDE and DDT, and studies have rarely investigated the effects of DDT or DDE exposure at levels observed in populations exposed through indoor residual spraying or populations exposed through drift of outdoor applications.

Rosas and Eskenazi (2008), in a review of the association between pesticides and neurodevelopment, concluded that although there are some inconsistencies among studies that may arise from differences in exposure and in methods of exposure assessment, there is surprising consistency in the few studies that have been conducted. The studies suggest that there is reason to be cautious about exposure of pregnant women to DDT, DDE, and organophosphates because of the potential effect on the neurodevelopment of their children.

Integrated Pest Management

If climate change causes the spread of diseases now considered to be tropical diseases into what are now more temperate climates, the use of pesticides could increase and have the potential to degrade indoor air quality. Alternatively, the concern about rodent and insect vectors could be used to promote wider use of Integrated Pest Management (IPM). IPM is an effective and environmentally sensitive approach to pest management that relies on a combination of common-sense practices. IPM programs use current comprehensive information on the life cycles of pests and their interactions with the environment. That information, in combination with available pest-control methods, is used to manage pest damage by the most economical means and with the least possible hazard to people, property, and the environment (EPA, 2010). IPM integrates common principles in the approach that should be taken to control pests in the environment. First is setting a threshold for action. For indoor environments, the potential of harming building integrity or human health is considered before any action. Families need to consider alternatives to chemical pesticides and avoid using the wrong kind of pesticide for a problem. Many intervention programs have been implemented to help families to recognize how to manage indoor space to prevent pests from posing a threat. The extent to which a family has the resources or control to modify its home environment will determine the success of its IPM interventions. If chemicals are needed, IPM assesses the proper control method for both effectiveness and risk minimization. Researchers are beginning to study the effectiveness of IPM interventions in high-risk populations. Williams et al. (2006) reported on an IPM intervention study of pregnant New York City black and Latina women; the study used education, sealing of pest entry points, and application of low-toxicity pesticides. They were able to show decreased cockroach infestation; lower indoor air concentrations of piperonyl butoxide, which is a synergist commonly added to pyrethroid insecticides that were applied; and lower levels of insecticides in maternal blood samples at delivery.

Synthesis

Generally speaking, alterations in outdoor environmental conditions may affect indoor exposures to pests by changing the habitable range of creatures known to invade indoor environments or by changing indoor environmental conditions or behavior in ways that drive them indoors. Buildings and building-maintenance practices that work well for one set of environmental conditions may not protect against infestations under other conditions. Termite infestations, for example, are less common in northern parts of the United States, and buildings and building codes there do not always require termite-prevention measures (Peterson, 2010). If termite ranges move northward, it may lead both to increased property damage and to occupant exposure to pesticides unless anticipatory maintenance and regulatory changes are made.

CONCLUSIONS

Several of the key findings of the 2001 National Research Council report *Under the Weather: Climate, Ecosystems, and Infectious Diseases* remain pertinent and bear repeating. They are excerpted and quoted below; additional explanatory detail is available in that report.

Key Findings Regarding Linkages Between Climate and Infectious Diseases from the Report *Under the Weather: Climate, Ecosystems, and Infectious Diseases*

- Weather fluctuations and seasonal-to-interannual climate variability influence many infectious diseases.
- Observational and modeling studies showing an association between climatic variations and disease incidence must be interpreted cautiously.
- Climate change may affect the evolution and emergence of infectious diseases.
- The relationships between climate and infectious disease are often highly dependent upon local-scale parameters and there are potential pitfalls in extrapolating climate and disease relationships from one spatial/temporal scale to another.
- The potential disease impacts of global climate change remain highly uncertain.

Research Needs and Surveillance Regarding Climate and Infectious Diseases from the Report *Under the Weather: Climate, Ecosystems, and Infectious Diseases*

- Research on the linkages between climate and infectious diseases must be strengthened.
- Further development of disease transmission models is needed to assess the risks posed by climatic and ecological changes.
- Epidemiological surveillance programs should be strengthened.
- Observational, experimental, and modeling activities are all highly interdependent and must progress in a coordinated fashion.
- Research on climate and infectious disease linkages inherently requires interdisciplinary collaboration.

Other Conclusions

In addition, on the basis of its review of the papers, reports, and other information presented in this chapter, the present committee has reached the following conclusions regarding infectious agents and pests:

- More investigation is needed to determine the extent to which the critical aspects of influenza spread are determined by indoor vs outdoor environmental conditions. It should consider air conditioning, which affects indoor temperature and humidity, and geographic location because there may be salient differences among regions in viral and human biology.
- The ecologic niches for house dust mites will change in response to climate change. Locations that are hotter and drier and that have increased use of air conditioning will tend to have fewer dust mite infestations. Decreased use of heating systems in winter because of milder conditions may result in increased dust mite populations.
- Decreases in dust mite populations in some locations may lower the incidence of allergic reactions to dust mites, but the overall incidence of allergic disease may not go down, because those who are predisposed to allergies may become sensitized to other air contaminants.
- Climate change may also lead to shifting patterns of indoor exposure to pesticides as occupants and building owners respond to infestations of pests whose ranges have changed.

REFERENCES

Acosta LM, Acevedo-Garcia D, Perzanowski MS, Mellins R, Rosenfeld L, Cortes D, Gelman A, Fagan JK, Bracero LA, Correa JC, Reardon AM, Chew GL. 2008. The New York City Puerto Rican asthma project: Study design, methods, and baseline results. *Journal of Asthma* 45(1):51–57.

Akinbami LJ. 2006. *The state of childhood asthma, United States, 1980–2005. Advance data from vital and health statistics* No. 381. U.S. Department of Health and Human Services, Centers for Disease Control and Prevention, National Center for Health Statistics.

Antens CJ, Oldenwening M, Wolse A, Gehring U, Smit HA, Aalberse RC, Kerkhof M, Gerritsen J, de Jongste JC, Brunekreef B. 2006. Repeated measurements of mite and pet allergen levels in house dust over a time period of 8 years. *Clinical and Experimental Allergy* 36(12):1525-1531.

Arbes SJ Jr, Gergen PJ, Elliott L, Zeldin DC. 2005. Prevalences of positive skin test responses to 10 common allergens in the US population: Results from the third National Health and Nutrition Examination Survey. *Journal of Allergy and Clinical Immunology* 116(2):377-383.

Arlian LG. 1975. Dehydration and survival of the European house dust mite, *Dermatophagoides pteronyssinus*. *Journal of Medical Entomology* 12(4):437-442.

Arlian LG, Veselica MM. 1981. Re-evaluation of the humidity requirements of the house dust mite *Dermatophagoides farinae* (Acari: Pyroglyphidae). *Journal of Medical Entomology* 18:351-352.

Arlian LG, Neal JS, Morgan MS, Vyszenski-Moher DL, Rapp CM, Alexander AK. 2001. Reducing relative humidity is a practical way to control dust mites and their allergens in homes in temperature climates. *Journal of Allergy and Clinical Immunology* 107(1):99-104.

Ayres JG, Forsberg B, Annesi-Maesano I, Dey R, Ebi KL, Helms PJ, et al. 2009. Climate change and respiratory disease: European Respiratory Society position statement. *The European Respiratory Journal* 34(2):295-302

Berkowitz GS, Obel J, Deych E, Lapinski R, Godbold J, Liu Z, et al. 2003. Exposure to indoor pesticides during pregnancy in a multiethnic, urban cohort. *Environmental Health Perspectives* 111(1):79-84.

Boner A, Pescollderungg L, Silverman M. 2002. The role of house dust mite elimination in the management of childhood asthma: An unresolved issue. *Allergy* 57(Suppl 74):23-31.

Boone SA, Gerba CP. 2007. Significance of fomites in the spread of respiratory and enteric viral disease. *Applied and Environmental Microbiology* 73(6):1687-1696.

Boxall AB, Hardy A, Beulke S, Boucard T, Burgin L, Falloon PD, et al. 2009. Impacts of climate change on indirect human exposure to pathogens and chemicals from agriculture. *Environmental Health Perspectives* 117(4):508-514.

Bradman A, Chevrier J, Tager I, Lipsett M, Sedgwick J, Macher J, et al. 2005. Association of housing disrepair indicators with cockroach and rodent infestations in a cohort of pregnant Latina women and their children. *Environmental Health Perspective* 113(12):1795-1801.

Brennon T. 2010. *Adaption and mitigation strategies for buildings in a changed climate.* Presentation at the National Academy of Sciences Workshop on Climate Change and Indoor Environment: Washington, DC.

Brunekreef B, van Strien R, Pronk A, Oldenwening M, de Jongst JC, Wijga A, Kerhof M, Aalberse RC. 2005. La mano di DIOS . . . was the PIAMA intervention study intervened upon? *Allergy* 60(8):1083-1086.

Carbrera P, Julià-Serdà G, Rodrgíuez de Castro F, Caminero J, Barber D, Carrillo T. 1995. Reduction of house dust mite allergens after dehumidifier use. *Journal of Allergy and Clinical Immunology* 95(2):635-636.
Carr W, Zeitel L, Weiss K. 1992. Variations in asthma hospitalizations and deaths in New York City. *American Journal of Public Health* 82:59-65.
CDC (Centers for Disease Control and Prevention). 2010a. Respiratory syncytial virus activity—United States, July 2008—December 2009. *MMWR* 59(8):230-233.
CDC. 2010b. Updated CDC estimates of 2009 H1N influenza cases, hospitalization and deaths in the United States, April 2009–April 10, 2010. http://www.cdc.gov/h1n1flu/estimates_2009_h1n1.htm (accessed February 2, 2011).
Chan ED, Iseman MD. 2010. Slender, older women appear to be more susceptible to nontuberculous mycobacterial lung disease. *Gender Medicine* 7(1):5-18.
Chew GL, Higgins KM, Gold DR, Muilenberg ML, Burge HA. 1999. Monthly measurements of indoor allergens and the influence of housing type in a northeastern US city. *Allergy* 54(10):1058-1066.
Chew GL, Perzanowski MS, Canfield SM, Goldstein IF, Mellins RB, Hoepner LA, et al. 2008. Cockroach allergen levels and associations with cockroach-specific IgE. *Journal of Allergy and Clinical Immunology* 121(1):240-245.
Chew GL, Reardon AM, Correa JC, Young M, Acosta L, Mellins R, et al. 2009. Mite sensitization among Latina women in New York, where dust-mite allergen levels are typically low. *Indoor Air* 19(3):193-197.
Cohn RD, Arbes SJ Jr, Yin M, Jaramillo R, Zeldin DC. 2004. National prevalence and exposure risk for mouse allergen in US households. *Journal of Allergy and Clinical Immunology* 113(6):1167-1171.
de Boer R, Kuller K. 1997. Mattresses as a winter refuge for house-dust mite populations. *Allergy* 52(3):299-305.
De Jong B, Zucs P. 2010. Legionella, springtime and potting soils. *Eurosurveillance* 15(8):2-3.
du Prel JB, Puppe W, Gröndahl B, Knuf M, Weigl JA, Schaaff F, Schmitt HJ. 2009. Are meteorological parameters associated with acute respiratory tract infections? *Clinical Infectious Diseases* 49(6):861-868.
Dulek DE, Peebles RS Jr. 2011. Viruses and asthma. *Biochimica et Biophysica Acta* doi:10.1016/j.bbagen.2011.01.012.
Eldeirawi K, McConnell R, Freels S, Persky VW. 2005. Associations of place of birth with asthma and wheezing in Mexican American children. *Journal of Allergy and Clinical Immunology* 116(1):42-48.
EPA (Environmental Protection Agency). 2010. Integrated pest management (IPM) strategies. http://www.epa.gov/pesticides/factsheets/ipm.htm (accessed February 3, 2011).
Eskenazi B, Bradman A, Castorina R. 1999. Exposures of children to organophosphate pesticides and their potential adverse health effects. *Environmental Health Perspectives* 107(Suppl 3):409-419.
Eskenazi B, Marks AR, Bradman A, Harley K, Barr DB, Johnson C. 2007. Organophosphate pesticide exposure and neurodevelopment in young Mexican-American children. *Environmental Health Perspectives* 115(5):792-798.
Eskenazi B, Rosas LG, Marks AR, Bradman A, Harley K, Holland N, et al. 2008. Pesticide toxicity and the developing brain. *Basic & Clinical Pharmacology & Toxicology* 102(2):228-236.
Eskenazi B, Chevrier J, Rosas LG, Anderson HA, Bornman MS, Bouwman H, et al. 2009. The Pine River statement: Human health consequences of DDT use. *Environmental Health Perspectives* 117(9):1359-1367.
Falkinham JO III, Iseman MD, de Haas P, van Soolingen D. 2008. Mycobacterium avium in a shower linked to pulmonary disease. *Journal of Water Health* 6(2):209-213.

Falkinham JO. 2010. Impact of human activities on the ecology of nontuberculous mycobacteria. *Future Microbiology* 5(6):951-960.

Feazel LM, Baumgartner LK, Peterson KL, Frank DN, Harris JK, Pace NR. 2009. Opportunistic pathogens enriched in showerhead biofilms. *Proceedings of the National Academy of Sciences of the United States of America* 106(38):16393-16399.

Fenske RA, Black KG, Elkner KP, Lee CL, Methner MM, Soto R. 1990. Potential exposure and health risks of infants following indoor residential pesticide applications. *American Journal of Public Health* 80(6):689-696.

Fisman DN, Lim S, Wellenius GA, Johnson C, Britz P, Gaskins M, Maher J, Mittleman MA, Spain CV, Haas CN, Newbern C. 2005. It's not the heat, it's the humidity: Wet weather increases legionellosis risk in the greater Philadelphia metropolitan area. *Journal of Infectious Diseases* 192(12):2066-2067.

Furlong CE, Holland N, Richter RJ, Bradman A, Ho A, Eskenazi B. 2006. PON1 status of farmworker mothers and children as a predictor of organophosphate sensitivity. *Pharmacogenetics and Genetics* 16(3):180-190.

Girman J. 2010. *Research needed to address the impacts of climate change on indoor air quality.* Washington DC: US Environmental Protection Agency.

Gehring U, Brunekreek B, Fahlbusch B, Wichmann HE, Heinrich J, INGA Study Group. 2005. Are house dust mite allergen levels influenced by cold winter weather? *Allergy* 60(8):1079-1082.

Grant EN, Wagener R, Weiss KB. 1999. Observations on emerging patterns of asthma in our society. *Journal of Allergy and Clinical Immunology* 104:S1-S9.

Gruchalla RS, Pongracic J, Plaut M, Evans R 3rd, Visness CM, Walter M, et al. 2005. Inner City Asthma Study: Relationships among sensitivity, allergen exposure, and asthma morbidity. *Journal of Allergy and Clinical Immunology* 115(3):478-485.

Gubler DJ, Reiter P, Ebi KL, Yap W, Nasci R, Patz JA. 2001. Climate variability and change in the United States: Potential impacts on vector- and rodent-borne diseases. *Environmental Health Perspectives* 109(Suppl 2):223-233.

Hales S, de Wet N, Maindonald J, Woodward A. 2002. Potential effect of population and climate changes on global distribution of dengue fever: An empirical model. *Lancet* 360(9336):830-834.

Harving H, Korsgaard J, Dahl R. 1994. Clinical efficacy of reduction in house-dust mite exposure in specially designed, mechanically ventilated "healthy" homes. *Allergy* 49(10): 866-870.

Holland SM, Vinh DC. 2009. Yeast infections–human genetics on the rise. *New England Journal of Medicine* 361(18):1798-1801.

Holt PG, Thomas WR. 2005. Sensitization to airborne environmental allergens: Unresolved issues. *Nature Immunology* 6(10):957-960.

Huss K, Adkinson NF Jr, Eggleston PA, Dawson C, Van Natta ML, Hamilton RG. 2001. House dust mite and cockroach exposure are strong risk factors for positive allergy skin test responses in the Childhood Asthma Management Program. *Journal of Allergy and Clinical Immunology* 107(1):48-54.

Ingram JM, Sporik R, Rose G, Honsinger R, Chapman MD, Platts-Mills TA. 1995. Quantitative assessment of exposure to dog (Can f 1) and cat (Fel d 1) allergens: Relation to sensitization and asthma among children living in Los Alamos, New Mexico. *Journal of Allergy and Clinical Immunology* 96(4):449-456.

Illi S, von Mutius E, Lau S, Niggemann B, Grüber C, Wahn U, et al. 2006. Perennial allergen sensitisation early in life and chronic asthma in children: A birth cohort study. *Lancet* 368(9537):763-770.

IOM (Institute of Medicine). 2000. *Clearing the air—Asthma and indoor air exposures*. Washington, DC: National Academy Press.

IOM. 2004. *Damp indoor spaces and health*. Washington, DC: The National Academies Press.
IOM. 2008a. *Global climate change and extreme weather events: Understanding the contributions to infectious disease emergence: Workshop summary*. Washington, DC: The National Academies Press.
IOM. 2008b. *Vector-borne diseases: Understanding the environmental, human health, and ecological connections: Workshop summary*. Washington, DC: The National Academies Press.
Johnston N, Sears M. 2006. Asthma exacerbations. 1: Epidemiology. *Thorax* 61:722-728.
Julien R, Levy JI, Adamkiewicz G, Hauser R, Spengler JD, Canales RA, et al. 2008. Pesticides in urban multiunit dwellings: Hazard identification using classification and regression tree (CART) analysis. *Journal of the Air and Waste Management Association* 58(10): 1297-1302.
Kattan M, Mitchell H, Eggleston P, Gergen P, Crain E, Redline S, et al. 1997. Characteristics of inner-city children with asthma: The National Cooperative Inner-City Asthma Study. *Pediatric Pulmonology* 24(4):253-262.
Lambert WE, Lasarev M, Muniz J, Scherer J, Rothlein J, Santana J, McCauley L. 2005. Variation in organophosphate pesticide metabolites in urine of children living in agricultural communities. *Environmental Health Perspectives* 113(4):504-508.
Lang DM, Polansky M. 1994. Patterns of asthma mortality in Philadelphia from 1969 to 1991. *New England Journal of Medicine* 331:1542-1546.
Lapeña S, Robles MB, Castañón L, Martínez JP, Reguero S, Alonso MP, Fernández I. 2005. Climatic factors and lower respiratory tract infection due to respiratory syncytial virus in hospitalised infants in northern Spain. *European Journal of Epidemiology* 20(3):271-276.
Lau S, Illi S, Sommerfeld C, Niggemann B, Bergmann R, von Mutius E, Wahn U. 2000. Early exposure to house-dust mite cat allergens and development of childhood asthma: A cohort study. Multicentre Allergy Study Group. *Lancet* 356(9239):1392-1397.
Lintner TJ, Brame KA. 1993. The effects of season, climate, and air-conditioning on the prevalence of Dermatophagoides mite allergens in household dust. *Journal of Allergy and Clinical Immunology* 91(4):862-867.
Lu C, Fenske RA, Simcox NJ, Kalman D. 2000. Pesticide exposure of children in an agricultural community: Evidence of household proximity to farmland and take home exposure pathways. *Environmental Health Research* 84(3):290-302.
Mannino DM, Homa DM, Akinbami LJ, Moorman JE, Gwynn C, Redd SC. 2002. Surveillance for asthma—United States, 1980–1999. *Morbidity and Mortality Weekly Report* 51(SS01):1-13.
Marder D, Targonski P, Orris P, Persky V, Addington W. 1992. Effect of racial and socioeconomic factors on asthma mortality in Chicago. *Chest* 101:4265-4295.
McCauley LA, Lasarey MR, Higgins G, Rothlein J, Muniz J, Ebbert C, Phillips J. 2001. Work characteristics and pesticide exposures among migrant agricultural families: A community-based research approach. *Environmental Health Perspectives* 109(5):533-538.
McDonald LC, Banerjee SN, Jarvis WR. 1999. Seasonal variation of Acinetobacter infections: 1987-1996. Nosocomial Infections Surveillance System. *Clinical Infectious Diseases* 29(5):1133-1137.
Meerhoff TJ, Paget JW, Kimpen JL, Schellevis F. 2009. Variation of respiratory syncytial virus and the relation with meteorological factors in different winter seasons. *The Pediatric Infectious Diseases Journal* 28(10):860-866.
Montealegre F, Sepulveda A, Bayona M, Quiñones C, Fernádez-Caldas E. 1997. Identification of the domestic mite fauna of Puerto Rico. *Puerto Rico Health Sciences Journal* 16(2):109-116.

Montealegre F, Fernández B, Delgado A, Fernández L, Román A, Chardón D, Rodríguez-Santana J, Medina V, Zavala D, Bayona M. 2004. Exposure levels of asthmatic children to allergens, endotoxins, and serine proteases in a tropical environment. *The Journal of Asthma* 41(4):485-496.

Morey PR. 2010. *Climate change and potential effects on microbial air quality in the built environment*. http://www.epa.gov/iaq/pdfs/climate_and_microbial_iaq.pdf (accessed February 3, 2011).

Morgan WJ, Crain EF, Gruchalla RS, O'Connor GT, Kattan M, Evans R 3rd, et al. 2004. Results of a home-based environmental intervention among urban children with asthma. *New England Journal of Medicine* 351(11):1068-1080.

Murdoch DR, Jennings LC. 2009. Association of respiratory virus activity and environmental factors with the incidence of invasive pneumococcal disease. *The Journal of Infection* 58(1):37-46.

Myatt TA, Johnston SL, Zuo Z, Wand M, Kebadze T, Rudnick S, Milton DK. 2004. Detection of airborne rhinovirus and its relation to outdoor air supply in office environments. *American Journal of Respiratory & Critical Care Medicine* 169(11):1187-1189.

Myatt TA, Kaufman MH, Allen JG, MacIntosh DL, Fabian MP, McDevitt JJ. 2010. Modeling the airborne survival of influenza virus in a residential setting: The impacts of home humidification. *Environmental Health* 9:55.

National Exposure Research Laboratory. 2005. Compendium of NERL-sponsored children's exposure data and tools for assessing aggregate exposure to residential-use pesticides in support of the August 2006 reassessment. FY2005 Annual Performance Measure 33. Research Triangle Park, North Carolina.

Naumova EN. 2006. Mystery of seasonality: Getting the rhythm of nature. *Journal of Public Health Policy* 27(1):2-12.

NRC (National Research Council). 1993. *Pesticides in the diets of infants and children*. Washington, DC: National Academy Press.

NRC. 2001. *Under the weather: Climate, ecosystems, and infectious diseases*. Washington, DC: National Academy Press.

Omer SB, Sutanto A, Sarwo H, Linehan M, Djelantik IG, Mercer D, Moniaga V, Moulton LH, Widjaya A, Muljati P, Gessner BD, Steinhoff MC. 2008. Climatic, temporal, and geographic characteristics of respiratory syncytial virus disease in a tropical island population. *Epidemiology and Infection* 136(10):1319-1327.

Patterson K, Strek ME. 2010. Allergic bronchopulmonary aspergillosis. *The Proceedings of the American Thoracic Society* 7(3):237-244.

Perencevich EN, McGregor JC, Shardell M, Furuno JP, Harris AD, Morris JG Jr, Fisman DN, Johnson JA. 2008. Summer peaks in the incidences of gram-negative bacterial infection among hospitalized patients. *Infection Control and Hospital Epidemiology* 29(12):1124-1131.

Perera FP, Rauh V, Tsai WY, Kinney P, Camann D, Barr D, et al. 2003. Effects of transplacental exposure to environmental pollutants on birth outcomes in a multiethnic population. *Environmental Health Perspectives* 111(2):201-205.

Peterson CJ. 2010. Termites and climate change: Here, there and everywhere? *EARTH Magazine* (January 2010):46-53.

Phipatanakul W, Eggleston PA, Wright EC, Wood RA. 2000a. Mouse allergen. I. The prevalence of mouse allergen in inner-city homes. The National Cooperative Inner-City Asthma Study. *Journal of Allergy and Clinical Immunology* 106(6):1070-1074.

Phipatanakul W, Eggleston PA, Wright EC, Wood RA; National Cooperative Inner-City Asthma Study. 2000b. Mouse allergen. II. The relationship of mouse allergen exposure to mouse sensitization and asthma morbidity in inner-city children with asthma. *Journal of Allergy and Clinical Immunology* 106(6):1075-1080.

Platts-Mills TA, Satinover SM, Naccara L, Litonjua AA, Phipatanakul W, Carter MC, Heymann PW, Woodfolk JA, Peters EJ, Gold DR. 2007. Prevalence and titer of IgE antibodies to mouse allergens. *Journal of Allergy and Clinical Immunology* 120(5):1058-1064.

Pongracic JA, Visness CM, Gruchalla RS, Evans R, Mitchell HE. 2008. Effect of mouse allergen and rodent environmental intervention on asthma in inner-city children. *Annals of Allergy, Asthma and Immunology* 101(1):35-41.

Prasad C, Hogan MB, Peele K, Wilson NW. 2009. Effect of evaporative coolers on skin test reactivity to dust mites and molds in a desert environment. *Allergy and Asthma Proceedings* 30(6):624-627.

Quarles W. 2007. Global warming means more pests. *The IPM Practitioner* XXIX(9/10):1-8.

Randolph SE, Rogers DJ. 2000. Fragile transmission cycles of tick-borne encephalitis virus may be disrupted by predicted climate change. *Proceedings. Biology Sciences/The Royal Society* 267(1454):1741-1444.

Rodman DM, Polis JM, Heltshe SL, Sontag MK, Chacon C, Rodman RV, Brayshaw SJ, Huitt GA, Iseman MD, Saavedra MT, Taussig LM, Wagener JS, Accurso FJ, Nick JA. 2005. Late diagnosis defines a unique population of long-term survivors of cystic fibrosis. *American Journal of Respiratory and Critical Care Medicine* 171(6):621-626.

Rosas LG, Eskenazi B. 2008. Pesticides and child neurodevelopment. *Current Opinions in Pediatrics* 20(2):191-197.

Schubert MS. 2009. Allergic fungal sinusitis: Pathophysiology, diagnosis and management. *Medical Mycology* 47(Suppl 1):S324-S330.

Shaman J, Kohn M. 2009. Absolute humidity modulates influenza survival, transmission, and seasonality. *Proceedings of the National Academy of Sciences of the United States of America* 106(9):3243-3248.

Shaman J, Goldstein E, Lipsitch M. 2010a. Absolute humidity and pandemic versus epidemic influenza. *American Journal of Epidemiology* 173(2):127-135.

Shaman J, Pitzer VE, Viboud C, Grenfell BT, Lipsitch M. 2010b. Absolute humidity and the seasonal onset of influenza in the continental United States. *PLoS Biology* 8(2):e100031.

Singleton RJ, Bulkow LR, Miernyk K, DeByle C, Pruitt L, Hummel KB, Bruden D, Englund JA, Anderson LJ, Lucher L, Holman RC, Hennessy TW. 2010. Viral respiratory infections in hospitalized and community control children in Alaska. *Journal of Medical Virology* 82(7):1282-1290.

Sloan C, Moore ML, Hartert T. 2011. Impact of pollution, climate, and sociodemographic factors on spatiotemporal dynamics of seasonal respiratory viruses. *Clinical and Translational Science* 4(1):48-54.

Stapleton F, Keay LJ, Sanfilippo PG, Katiyar S, Edwards KP, Naduvilath T. 2007. Relationship between climate, disease severity, and causative organism for contact lens-associated microbial keratitis in Australia. *American Journal of Ophthalmology* 144(5):690-698.

Stern DA, Lohman IC, Wright AL, Taussig LM, Martinez FD, Halonen M. 2004. Dynamic changes in sensitization to specific aeroallergens in children raised in a desert environment. *Clinical and Experimental Allergy* 34(10):1563-1669.

Stevenson LA, Gergen PJ, Hoover DR, Rosenstreich D, Mannino DM, Matte TD. 2001. Sociodemographic correlates of indoor allergen sensitivity among United States children. *Journal of Allergy and Clinical Immunology* 108(5):747-752.

Stout JE, Yu VL. 1997. Legionellosis. *New England Journal of Medicine* 337(10):682-687.

Stout DM 2nd, Bradham KD, Egeghy PP, Jones PA, Croghan CW, Ashley PA, et al. 2009 American Healthy Homes Survey: A national study of residential pesticides measured from floor wipes. *Environmental Science and Technology* 43(12):4294-4300.

Sundell J, Wickman M, Pershagen G, Nordvall SL. 1995. Ventilation in homes infested by house-dust mites. *Allergy* 50(2):106-112.

Surgan MH, Congdon T, Primi C, Lamster S, Louis-Jacques J. 2002. Pest control in public housing, schools and parks: Urban children at risk. LAW 180-4 PESP 202-7643. Albany, NY: New York State Department of Law, Environmental Protection Bureau.

Talbot TR, Pehling KA, Hartert TV, Arbogast PG, Halasa NB, Edwards KM, et al. 2005. Seasonality of invasive pneumococcal disease: Temporal relation to documented influenza and respiratory syncytial viral cirvulation. *American Journal of Medicine* 118(3):285-291.

Tang JW, Lai FY, Nymadawa P, Deng YM, Ratnamohan M, Petric M, Loh TP, Tee NW, Dwyer DE, Barr IG, Wong FY. 2010a. Comparison of the incidence of influenza in relation to climate factors during 2000-2007 in five countries. *Journal of Medical Virology* 82(11):1958-1965.

Tang JW, Lai FY, Wong F, Hon KL. 2010b. Incidence of common respiratory viral infections related to climate factors in hospitalized children in Hong Kong. *Epidemiology and Infection* 138(2):226-235.

Tepas EC, Litonjua AA, Celedón JC, Sredl D, Gold DR. 2006. Sensitization to aeroallergens and airway hyperresponsiveness at 7 years of age. *Chest* 129(6):1500-1508.

Thier A, Enck J, Klossner C. 1998. *Plagued by pesticides: An analysis of New York State's 1997 pesticide use and sales data.* Albany, NY: Environmental advocates.

Tovey ER, Rawlinson WD. 2011. A modern miasma hypothesis and back-to-school asthma exacerbations. *Medical Hypotheses* 76(1):113-116.

Tsiodras S, Samonis G, Boumpas DT, Kontoyiannis DP. 2008. Fungal infections complicating tumor necrosis factor alpha blockade therapy. *Mayo Clinic Proceedings* 83(2):181-194.

Van Strien RT, Verhoeff AP, Brunekreef B, Van Wijnen JH. 1994. Mite antigen in house dust: Relationship with different housing characteristics in The Netherlands. *Clinical and Experimental Allergy* 24(9):843-853.

Voorhorst R, Spieksma FThM, Varekamp N. 1969. House dust mite atopy and the house dust mite *Dermatophagoides pteronyssinus* (Troussart, 1897). Leiden: Stafleu's Scientific Publishing Co.

Ward MH, Lubin J, Giglierano J, Colt JS, Wolter C, Bekiroglu N, et al. 2006. Proximity to crops and residential exposure to agricultural herbicides in Iowa. *Environmental Health Perspectives* 114(6):893-897.

Watson M, Gilmour R, Menzies R, Ferson M, McIntyre P; New South Wales Pneumococcal Network. 2006. The association of respiratory viruses, temperature, and other climatic parameters with the incidence of invasive pneumococcal disease in Sydney, Australia. *Clinical Infectious Diseases* 42(2):211-215.

Weiss KB, Wagner DK. 1990. Changing patterns in US asthma mortality: Identifying populations at high risk. *Journal of the American Medical Association* 264:1683-1687.

Welliver R. 2009. The relationship of meteorological conditions to the epidemic activity of respiratory syncytial virus. *Pediatric Respiratory Reviews* 10(Suppl 1):6-8.

Williams MK, Barr DB, Camann DE, Cruz LA, Carlton EJ, Borjas M, Reyes A, Evans D, Kinney PL, Whitehead RD, Jr., Perera FP, Matsoanne S, Whyatt RM. 2006 An intervention to reduce residential insecticide exposure during pregnancy among an inner-city cohort. *Environmental Health Perspectives* 114(11):1684-1689.

Wood JP, Choi YW, Chappie DJ, Rogers JV, Kaye JZ. 2010. Environmental persistence of a highly pathogenic avian influenza (H5N1) virus. *Environmental Science & Technology* 44(19):7515-7520.

WHO (World Health Organization). 2007. *WHO guidelines for indoor air quality: Dampness and mould.* Denmark: WHO Regional Office for Europe.

WHO. 2009. *WHO and DDT for malaria control—June 2009: WHO position statement.* Geneva: WHO Press.

Whyatt RM, Camann DE, Kinney PL, Reyes A, Ramirez J, Dietrich J, et al. 2002. Residential pesticide use during pregnancy among a cohort of urban minority women. *Environmental Health Perspectives* 110(5):507-514.

Whyatt RM, Barr DB, Camann DE, Kinney PL, Barr JR, Andrews HF, et al. 2003. Contemporary-use pesticides in personal air samples during pregnancy and blood samples at delivery among urban minority mothers and newborns. *Environmental Health Perspectives* 111(5):749-756.

Yamamoto Y, Nakamura K, Yamada M, Mase M. 2010. Persistence of avian influenza virus (H5N1) in feathers detached from bodies of infected domestic ducks. *Applied and Environmental Microbiology* 76(16):5496-5499.

Zoumot Z, Wilson R. 2010. Respiratory infection in noncystic fibrosis bronchiectasis. *Current Opinion in Infectious Diseases* 23(2):165-170.

Zwiener RJ, Ginsburg CM. 1988. Organophosphate and carbamate poisoning in infants and children. *Pediatrics* 81(1):121-126.

7

Thermal Stress

INTRODUCTION

This chapter addresses problems of indoor environmental quality associated with the thermal environment of buildings, how climate change could induce alterations in the frequency or severity of problems, and some of the means available to mitigate adverse conditions. Thermal stress is a particular threat to certain populations whose health, economic situation, or social circumstances make them vulnerable to exposure to temperature extremes or the consequences of such exposure. The text thus focuses its discussion of health effects on these vulnerable populations.

National Academies reports note that the first decade of the 21st century was 0.8°C (1.4°F) warmer than the first decade of the 20th century (NRC, 2010). Associated with that temperature rise have been observations that heat waves have become longer and more extreme and that cold spells have become shorter and milder. Because climate models suggest that those trends will continue and intensify, much of the information presented in the chapter relates to issues involving prolonged exposure to high temperature.

The climate change research that the committee relied on is summarized in Chapter 2. Studies of building ventilation—which plays a large role in determining indoor thermal conditions—are addressed in Chapter 8.

MANAGEMENT OF THE INDOOR THERMAL ENVIRONMENT

Buildings must protect occupants against extremes in outdoor temperatures. This section addresses the management of the indoor thermal environment, focusing on amelioration of high or prolonged heat conditions. Temperature fluctuations and prolonged exposure to low temperatures may also have health consequences. Generally, warmer conditions may lower the risk of health consequences among segments of the population that have difficulty in paying for heating during winter (Curriero et al., 2002; McGeehin and Mirabelli, 2001), but it should be noted that this benefit might be offset by circumstances in which weather extremes result in the loss of power for extended periods (MMWR, 1998).

Thermal Comfort Indoors

The American Society of Heating, Refrigerating and Air-Conditioning Engineers (ASHRAE) defines human thermal comfort as "the state of mind that expresses satisfaction with the surrounding environment" (ASHRAE, 2004). Although comfort is a subjective evaluation, survival and health are affected by temperature, humidity, and individual factors (such as clothing, air speed, metabolic rate, and health) related to the generation, dissipation, and retention of body heat. In addition to outdoor temperature, humidity, and solar radiation, comfort is influenced by whether a building has air conditioning and whether occupants have control over the temperature (Nicol and Humphreys, 2002). Acclimatization plays a role; people who live in areas where high heat and humidity are common are better able to tolerate such conditions than those who do not (de Dear and Brager, 1998). And thermal comfort is influenced by radiant heat transfer from surrounding objects: people near hot or cold surfaces feel warmer or cooler independently of the air temperature (EPA, 2009b).

"Typical" indoor temperature varies by season, locale, building type, and the economic circumstances of the occupants, although commercial spaces, such as offices, are often maintained at a more consistent year-round temperature than residences. ASHRAE's *Thermal Environmental Conditions for Human Occupancy* Standard 55-2004 characterizes the indoor summer comfort range[1] as about 74–83°F (23–28°C) and the winter comfort range[2] as about 67–79°F (19–26°C), depending on the relative humidity. ASHRAE separately defines acceptable temperature ranges for naturally ventilated spaces as a function of outdoor temperatures spanning about 50–93°F (10–34°C).

[1] More specifically, the range when occupants are dressed in clothing typically "worn when the outdoor environment is warm" (ASHRAE, 2004).

[2] When occupants are dressed in clothing typically worn when the outdoor environment is cool.

Effects of Climate Change on the Indoor Thermal Environment

Little research has addressed specifically the potential effects of climate change on the indoor thermal environment. The major issues surrounding this topic and some information addressing it are outlined below.

Indoor temperature is a function of outdoor temperature, the amount of solar radiation striking the structure, building insulation and ventilation characteristics, factors that influence the ability of the structure to dissipate stored heat, intentional sources of heat (heating, ventilating, and air-conditioning [HVAC] systems), and other indoor sources of heat (artificial lighting, cooking appliances, occupant metabolic heat, and the like). Scott and Huang (2007) found that the demand for cooling energy increases by 5–20% for every 1°C (1.8°F) increase in outdoor temperature, depending on the assumptions used.[3] Greater use of air conditioning for cooling implies more electricity demand, which is likely (at least in the short term) to be met through heavier use of fossil fuels, including coal, which in turn may lead to higher emissions of air pollutants, including the greenhouse gases that have been implicated in increased outdoor temperatures (IPCC, 2007). The positive feedback loop that characterizes those relationships is depicted in Figure 7-1.

The US Climate Change Science Program's literature review concluded that "temperature increases with global warming would increase peak demand for electricity in most regions of the country" but that research results varied and were influenced by such factors as "whether the study allows for changes in the building stock and increased market penetration of air conditioning in response to warmer conditions" (Scott and Huang, 2007). Indoor relative humidity, another component of the thermal environment, is a part of the issue. In areas of the country where hot and humid outdoor conditions become more common, air-conditioning units may run longer to restore or maintain comfortable indoor humidity.

Potential increases in the magnitude and frequency of peak electricity demand due to heat waves and in the occurrence of extreme weather events have also led to concerns over power outages that could leave building occupants without sources of conditioned air. The 1995 Chicago (Changnon et al., 1996) and 1999 New York City (USGCRP, 2009) heat waves were accompanied by extended and widespread power outages. Electric-grid infrastructure disruptions after Hurricanes Katrina and Rita left some areas of the southern United States without power for weeks during the late summer of 2005.

---

[3] The same study found that demand for heating energy decreases by 3–15% for every 1°C (1.8°F) increase in outdoor temperature. Cooling uses electricity almost exclusively whereas heating uses various energy sources; this complicates the evaluation of the implications of these changes on overall power-generation demands.

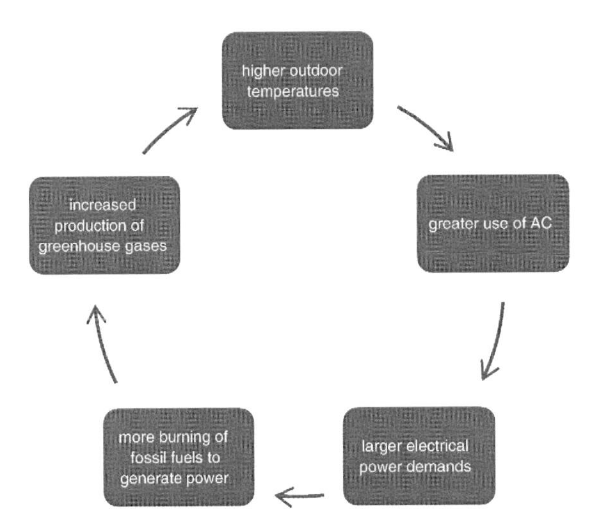

FIGURE 7-1 The relationship between outdoor temperature, air-conditioning use, electric-power demand, and greenhouse-gas generation.

EFFECTS OF HEAT EXPOSURE

Healthy people can physically adapt to changes in ambient temperature within some limits. However, when temperatures push the upper end of those limits or are combined with other factors—such as high humidity, strenuous activity, or prolonged exposure—physiologic compensation mechanisms can be overwhelmed. The National Weather Service's (NWS's) Heat Index—a measure of perceived temperature derived from the ambient temperature and relative humidity and based on work originally conducted by Steadman (1979)—is an imperfect but useful tool in determining potential health threats (Metzger et al., 2010). Figure 7-2 illustrates heat-index values for a range of temperature and humidity combinations and indicates the corresponding NWS health-threat level.

A 2011 review by Anderson and Bell examined the determinants of mortality in heat waves through an empirical analysis of 43 events in US cities over the years 1987–1995. Mortality increased an average of 3.74% (95% Confidence Interval [CI] 2.29–5.22%) on heat wave days versus non-heat wave days. The largest effect was observed in the Northeast and Midwest US census regions, the smallest in the South, even though the longest heat waves occurred in that region. Analyses also found that heat waves at the beginning of the warm weather months had greater mortality effects (5.04%, nationally) than those later in the season (2.65%). The investigators speculated that these results were due to behavioral and physiological acclimatization.

Temperature (°F)

Relative Humidity (%)	80	82	84	86	88	90	92	94	96	98	100	102	104	106	108	110
40	80	81	83	85	88	91	94	97	101	105	109	114	119	124	130	136
45	80	82	84	87	89	93	95	100	104	109	114	119	124	130	137	
50	81	83	85	88	91	95	99	103	108	113	118	124	131	137		
55	81	84	86	89	93	97	101	106	112	117	124	131	137			
60	82	84	88	91	95	100	105	110	116	123	129	137				
65	82	85	89	93	98	103	108	114	121	128	136					
70	83	86	90	95	100	105	112	119	126	134						
75	84	88	92	97	103	109	116	124	132							
80	84	89	94	100	106	113	121	129								
85	85	90	96	102	110	117	126	135								
90	86	91	98	105	113	122	131									
95	86	93	100	108	117	127										
100	87	95	103	112	121	132										

Likelihood of Heat Disorders with Prolonged Exposure or Strenuous Activity

Caution | Extreme Caution | Danger | Extreme Danger

FIGURE 7-2 National Weather Service Heat-Index values and corresponding health-threat levels (NWS, 2010).

Physiologic Vulnerability to Heat Events

A number of biological factors influence the ability of people to adapt to high temperature conditions or withstand extended exposure to them. These factors are identified and discussed below.

As people age, their ability to cope with external environmental stressors decreases. That is based on both physiologic and social factors: decreased organ function, interactions between medications and heat-compensation mechanisms, overall poor health status, isolation, and decreased access to support services.

There are stark physiologic differences between younger adult and elderly populations. Decreased organ function is a major issue. The peripheral nervous system is affected by the aging process: myelin sheaths deteriorate, and myelinated and unmyelinated nerve fibers are lost. The peripheral nervous system tells the body to feel hot and cold. It also regulates internal processes, such as heart rate and contraction and expansion of blood vessels, to maintain proper blood pressure and the body's reaction to stress. Decreased sensation may limit a person's ability to recognize that she or he needs to take steps to decrease body temperature. Sweat production and sweat-gland functioning, which are coping mechanisms to reduce the body's core temperature, are also regulated by the peripheral nervous system. The number of sweat glands does not decrease with age, but sweat production does, and this makes it difficult to reduce the body's core temperature (Verdú et al., 2000).

The overall health status of the elderly is poorer than that of other age groups. The elderly exhibit higher rates of chronic ailments, including cardiovascular diseases, diabetes, chronic obstructive pulmonary disease, diabetes, renal disease, and neoplasms (Khalaj et al., 2010; Pearlman and Uhlmann, 1988; Reid et al., 2009). Cardiovascular disease has been identified as the most important risk factor for heat stroke in the elderly (Kenney and Munca, 2003), but other chronic illnesses, such as those mentioned above, are also known to increase the risk of heat stroke (Khalaj et al., 2010).

Some medications, including over-the-counter supplements, may have adverse thermoregulatory effects. Psychotropic drugs have been associated with a higher risk of hospitalization of the elderly due to hyperthermia (Lopez and Goldoftas, 2009). Nonsteroidal anti-inflammatory drugs, such as aspirin—which is commonly taken for myocardial-infarction prevention—block prostaglandins, which aid in controlling body temperature and blood pressure (Carmichael and Shankel, 1985). Anticholinergics inhibit sweat production; younger persons also use these medications, but their sweating process is not affected, changes having been noted only in those who were about 80 years old or older (Kenney and Munca, 2003). Other medications, such as diuretics, limit cutaneous vasodilation and pose a high risk of dehydration, a particular concern during heat stress (Kenney and Munca, 2003).

Those suffering from chronic diseases are also at risk. Research indicates that obesity, hypertension, diabetes, and cardiovascular disease increase susceptibility to the effects of extreme heat.

Obesity is a recognized public-health concern. Few studies have looked specifically at obese or overweight persons and heat waves, but some information is available. Obesity was a comorbidity in the 2003 European heat wave (Vandentorren et al., 2006); this is not surprising given that fatal heat strokes occur at a rate 3.5 times higher in those who are obese or overweight than in those of normal weight (Kenny et al., 2010). That may be because of a lowered capacity of heat dissipation due to a low ratio of body surface area to body mass, which hinders sweat evaporation (Kenny et al., 2010). Adipose tissue also stores heat more efficiently than other tissues, such as muscle, and subcutaneous fat restricts conductive heat transfer (Kenny et al., 2010).

According to the Centers for Disease Control and Prevention, the prevalence of hypertension is about 30% in the United States (Fryar et al.,

2010). A study of the elderly in Baltimore, Maryland, found that 50% of those who experienced adverse heat symptoms during the summer months had a history of hypertension (Basu and Samet, 2002). Hypertension was a common comorbidity factor in those who died from heat effects during the Chicago 1995 heat wave (Dematte et al., 1998). Impairments of circulation, such as those which occur in people who have hypertension, may reduce blood flow to the dermis, and this may weaken temperature regulation by reducing heat transfer from the core to the skin (Carberry et al., 1992; Kenny et al., 2010).

Diabetes occurs in about 10% of the US population (Fryar et al., 2010), and studies have shown that those who have diabetes suffer disproportionately during extreme heat events compared with the general population (Kenny et al., 2010). Circulatory changes, such as vessel dilation and vascular reactivity, are greatly compromised in those who have diabetes (Kenny et al., 2010; Petrofsky et al., 2005; Stansberry et al., 1997). Neuropathy, which is common in diabetic people, impedes sweat responses (Fealy et al., 1989; Kenny et al., 2010). Diabetic people also may have fluid and electrolyte disturbances, which affect glucose regulation (Semenza, 1999); this was seen in a heat wave in New York and St. Louis in 1966, where those who had diabetes had increased mortality (Schuman, 1966).

Cardiovascular diseases afflict about 12% of Americans (CDC, 2010). Although there are few studies of cardiovascular disease and heat, some links have been found between increased mortality during heat waves and the presence of cardiovascular diseases (Hoffmann et al., 2008; Kenny et al. 2010; Klinenberg, 2002). Like other diseases that disrupt cardiovascular flow, cardiovascular diseases impair body-temperature regulation. Mortality in those who had cardiovascular diseases was 30% higher during the 2003 European heat wave than during other "normal" heat days (Hoffmann et al., 2008). Cardiovascular disease was prominent among the chronic diseases blamed for the excess mortality in France during the 2003 heat event (Fouillet et al., 2006; Vandentorren et al., 2006), and the same was observed during the 1995 Chicago heat wave (Klinenberg, 2002).

Economic and Social Vulnerability to Heat Events

Several studies have examined how economic and social circumstances influence vulnerability to death and disease associated with heat-wave events. Shonkoff and colleagues (2009) published a review of the literature focused on the disparate effects of climate change in California on groups of lower socioeconomic status. Heat waves in that state and others resulted in increased emergency-department visits for acute renal failure, diabetes, cardiovascular disease, electrolyte imbalance, and nephritis (Knowlton et al., 2009; Kovats and Hajat, 2008). Children 4 years old and younger and people over 65 years old were at greatest risk. Other investigators have found that low-income black Americans are disproportionately affected (Basu and Ostro, 2008; Medina-Ramon et al., 2006; O'Neill et al., 2003). Analysis has shown that it is unlikely that this was a result of racial differences in physiology but rather a consequence of lower socioeconomic status, the physical settings that they live in, and their greater exposure to high temperatures (Basu and Ostro, 2008).

The poor are more likely to be living in homes that do not have air conditioning. According to the American Housing Survey (AHS), about half of those living below the national poverty line do not have air conditioning in their homes (USCB, 2009). The elderly may lack the financial resources to make the necessary modifications to adapt to the heat, such as installing air-conditioning units. Low socioeconomic status also has more subtle effects. Those living in lower-income areas may experience higher rates of crime. In the Chicago 1995 heat wave, some elderly people restricted ventilation in their homes by not opening windows for fear of crime (Klinenberg, 2002). Fear of crime leads people to stay in their homes, and this increases mortality in heat events (Klinenberg, 2002; Lopez and Goldoftas, 2009). People of lower socioeconomic status who have chronic health problems are disproportionately affected by medical conditions because of their lack of access to care and of the resources needed to manage their diseases effectively (Phelan et al., 2004). People of low socioeconomic status who belong to some minority groups are also less likely to have access to private transportation, so their ability to move to community sites that have air conditioning is restricted. Disparities in air-conditioning access contributed to the difference in heat-wave mortality, which was nearly twice as high in minority-group residents in Los Angeles as the average in Los Angeles (Kovats and Hajat, 2008).

Social isolation is a large factor in predicting heat morbidity, particularly among the elderly. According to the US Census Bureau, about 25% of the general population and 32% of the elderly population live alone (Klinenberg, 2002). Physical impairments and mobility restriction due to age and other limitations may prevent people—particularly those who live on upper floors—from leaving their home and reaching cooling centers set up by the community (Lopez and Goldoftas, 2009). In the 1995 Chicago heat wave, several trends due to social isolation were discovered. For example, 73% of heat-related deaths were in people over 65 years old, and those who lived alone were at additional risk for death (Klinenberg, 2002; Semenza et al., 1996); and those who did not leave their homes at least once a day and did not have access to transportation had higher mortality (Semenza et al., 1996). Similar trends were found in the 1999 Chicago heat wave (Naughton et al., 2002).

The so-called heat-island effect may also be a factor in higher heat-related morbidity and mortality found in urban areas than in rural areas (Hajat et al., 2007; Martinez et al., 1989). It involves circumstances in which urban areas are hotter than surrounding rural areas because of the presence of large numbers of buildings, parking lots, and other infrastructure that has a great ability to store solar energy (Basu and Samet, 2002; Luber and McGeehin, 2009). It is more common in locales that have relatively few green spaces. A heat island absorbs and stores heat during the day and radiates it during the night, sustaining higher temperatures and intensifying the effects of heat waves (Luber and McGeehin, 2009). Green spaces are associated with decreased heat-related morbidity and mortality that are due to heat-island effects and the overall lack of direct shading for residents (Kilbourne et al., 1982; Reid et al., 2009; Tan et al., 2007).

Shonkoff and colleagues' (2009) review paper describes an unpublished analysis by Morello-Frosch and Jesdale (2008), who found a positive dose–response relationship between the presence of impervious surfaces and high community poverty and a negative dose–response relationship between the amount of tree cover and the extent of community poverty in four California urban areas. That suggested the potential for a greater burden of heat-island exposure of low-income populations than of higher-income populations. The relationship was also observed by researchers in Phoenix, Arizona, who found that elderly, minority-group, and low-income residents were at the highest risk for exposure to extreme heat (Ruddell et al., 2010).

The lack of access to air conditioning thus directly influences the risk of high heat exposure and heat-related morbidity and mortality. It also plays an important role in home ventilation, which affects exposure to air pollutants and overall indoor air quality apart from temperature.

Air-Conditioning Prevalence and Use

Air conditioning has been the primary means of moderating high temperatures in buildings in the United States since the 1950s. The fraction of homes in the United States that have air conditioning has risen steadily over the past 40 years, from 46.9% of year-round units[4] in 1973 to 87.4% in 2005 (Eggers and Thackeray, 2007). The type of air-conditioning unit has shifted over that time. In homes, central air-conditioning systems[5] were present in 16.8% of year-round units in 1973, to 33.2% in 1985, 47.0% in 1995, and 65.4% in 2005. Only 12.6% of year-round units were without

[4] Year-round units are defined by the Census Bureau as "those intended for occupancy at any time of the year, even though they may not be in use the year round" (USCB, 2004).

[5] A central air-conditioning system is one that "uses ducts to distribute cooled and/or dehumidified air to more than one room or uses pipes to distribute chilled water to heat exchangers in more than one room, and which is not plugged into an electrical convenience outlet" (266 CMR 2.00 Definitions, Massachusetts Office of Consumer Affairs & Business Regulation).

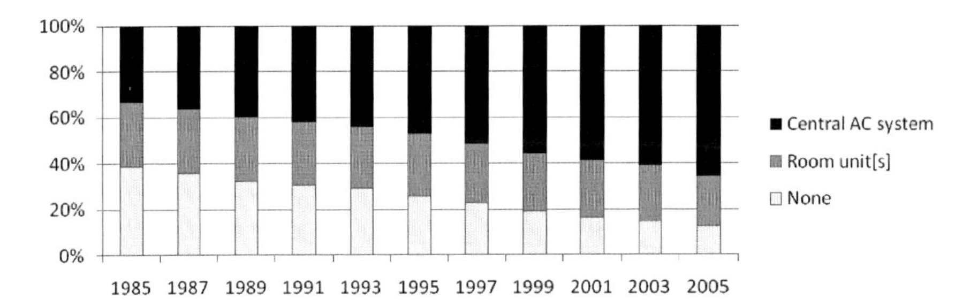

FIGURE 7-3 Percentages of year-round units in the United States with central air-conditioning systems, one or more room units, or no air conditioning, 1985–2005 (Eggers and Thackeray, 2007, derived from American Housing Survey data).

any form of air conditioning by 2005. Figure 7-3 illustrates changes in the prevalence and type of air conditioning in residences over the past 25 years.

There are substantial variations in air-conditioning system prevalence in different parts of the country. AHS data for 2005 indicate, unsurprisingly, that air-conditioning is more common in the southern and southwestern United States than elsewhere[6] and in the parts of the country that typically have the most cooling degree days and the fewest heating degree days.[7] Figure 7-4 details those data.

The climate zone and census region that encompass California exhibit relatively lower penetration of air-conditioning units than might be expected. Many homes in California are not equipped with air conditioning, because coastal temperatures are relatively mild during summer (Basu and Ostro, 2008). The reduced use of air-conditioning equipment is also influenced by the state energy and efficiency programs that include "cool community" standards for shading (Brown and Koomey, 2003).

In addition to the increase in air-conditioning units, the hours during which air conditioning is used have increased over the years. The Depart-

[6] The southern and southwestern parts of the United States were experiencing rapid growth in new construction at this time, and this accounts in part for the greater prevalence of air conditioning.

[7] Cooling degree days are used to estimate how hot the climate is and how much energy may be needed to keep buildings cool. Cooling degree days are calculated by subtracting a balance temperature from the mean daily temperature and summing only positive values over an entire year. Heating degree days are used to estimate how cold the climate is and how much energy may be needed to keep buildings warm. Heating degree days are calculated by subtracting the mean daily temperature from a balance temperature and summing only positive values over an entire year. The balance temperature used can vary but is usually set at 65°F (18°C), 68°F (20°C), or 70°F (21°C) (EPA, 2009b).

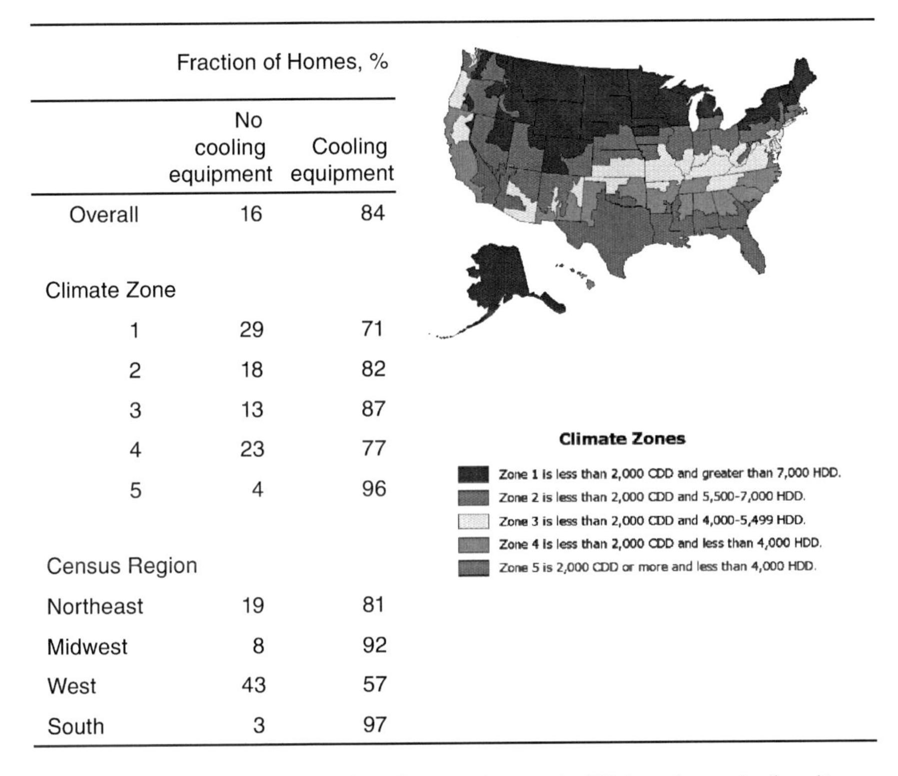

	Fraction of Homes, %	
	No cooling equipment	Cooling equipment
Overall	16	84
Climate Zone		
1	29	71
2	18	82
3	13	87
4	23	77
5	4	96
Census Region		
Northeast	19	81
Midwest	8	92
West	43	57
South	3	97

FIGURE 7-4 Percentage use of cooling equipment in US housing units by climate zone and census region, 2005 (EIA, 2010a,b for data; EIA, 2007 for figure). NOTE: The data used in this table differ from those used to generate Figure 7-3, which were based on year-round units only.

ment of Energy's Residential Energy Consumption Survey found that 33% of residences that had central air conditioning and 11% of residences that had window or wall units reported using an air conditioner "all summer" in 1981 (DOE, 2000). By 1997, those figures had risen to 52% and 21%, respectively (DOE, 2000), and in 2005, 61% and 30% (DOE, 2008). Collectively, 73% of residences that had any form of air conditioning reported using it either "all summer" or "quite a bit" in 2005 (DOE, 2008).

Most central air conditioners in residences have no outside air intakes, unlike the window and wall units that they sometimes supplanted. Instead, they rely on the infiltration of outdoor air through windows and doors and on loose construction. Climate change may stimulate the implementation of energy-efficiency (also called weatherization) measures that limit such infiltration and may lead to inadequate ventilation, as discussed in Chapter 8.

Most new commercial buildings have central air conditioning and mechanical ventilation (US Energy Information Administration, 2006), but the details of HVAC design and energy-use considerations vary considerably by the type (such as office and retail), size, age, and location of the structure. The presence of air conditioners in schools depends on several factors, especially geographic location. A 2005 US Department of Education survey of public-school principals found that 31% of permanent school buildings in the Northeast, 25% in the central region, 14% in the West, and 1% in the South did not have air conditioning (ED, 2007).

Epidemiologic Research on Effects of Air Conditioning on Health and Productivity

There is a small literature that examines occupant health and productivity in buildings that have air-conditioning systems vs buildings that rely on natural ventilation. Mendell and colleagues (2008) note that "the presence of central mechanical air-conditioning systems in office buildings (relative to natural ventilation) is one of the risk factors associated most consistently with increased" sick-building syndrome symptoms. In a 2004 multiple-building study of middle-aged "professional" women in France, Preziosi et al. reported that sickness absence, medical-services use (doctor visits), and hospital stays were 57%, 17%, and 35% fewer, respectively, among subjects who had natural ventilation in their workplaces than among those who had air conditioning. Hummelgaard and colleagues (2007) reported that reported building-related health symptoms were 31% fewer among occupants of nine naturally ventilated office buildings in Denmark than among occupants of mechanically ventilated offices. Sahakian et al. (2009) found that office workers who had home air conditioning were more likely to have visited a medical specialist in the previous year (prevalence ratio, 1.3; P = 0.02) than those who had naturally ventilated homes. The reasons for those outcomes are not clear and might vary by study.

A 1996 study by Aldous and colleagues found that infants exposed to home evaporative cooling systems experienced a higher risk of wheezing lower respiratory illness than those in homes without such systems (odds ratio = 1.8; 95% CI, 1.1–3.0). The authors speculated that the increased indoor humidity caused by evaporative cooling might support fungi or dust mites and associated adverse exposures. Evaporative coolers also increase the exchange of indoor with outdoor air and thus increase the levels of outdoor pollutants indoors, a concern if climate change results in higher outdoor pollutant levels.

Several potential explanations for those health outcomes have been put forward. Mendell et al. (2008) list exposures to microorganisms growing on wet surfaces in HVAC systems (in cooling coils, drain pans, and humidifiers), chemical biocides used in some humidifier systems, and poor HVAC maintenance in general among the possibilities but note that research to test these hypotheses was lacking.

In contrast, research suggests that air conditioning may provide protection against air pollutants of outdoor origin. Bell and Dominici (2008) analyzed data on 98 urban communities derived from the National Morbidity, Mortality, and Air Pollution Study, the US Census, and the AHS. They found that an increase in households that had central air conditioning was associated with a decrease in estimates of ozone's effect on mortality, but they cautioned that it was difficult to determine the extent to which that association was related to the presence of central air conditioning rather than something else. Bell et al. (2009) found that communities with higher percentages of households that had air conditioning had lower short-term effects of particulate matter ($PM_{2.5}$) on cardiovascular hospital admissions and that the effect was greater in connection with central air conditioning than with other forms, such as window units.

A conference paper by Fisk and Seppänen (2007) summarized the results of studies of the association between temperature and productivity measures in office and school environments. They found that controlling indoor temperatures in summer and ensuring adequate ventilation rates were associated with improved work and school performance. Lan et al. (2010) drew similar conclusions in their laboratory study of performance on a variety of calculation, learning, office-support, and reasoning tasks. They cautioned, though, that the extent to which experimental studies like theirs applied to actual office environments was not clear. Cost–benefit analyses conducted by Fisk and Seppänen (2007) suggested that "measures to improve indoor temperature control and increase ventilation rates will be highly cost effective, with benefit–cost ratios as high as 80 and annual economic benefits as high as $700 per person."

Chapter 8 summarizes the epidemiologic literature on a related issue, the health and productivity effects of ventilation.

EFFECTS OF COLD EXPOSURE

As Chapter 2 notes, measurements of global mean temperature in recent years indicate that cold spells in the United States have become shorter and milder. Some researchers have speculated that this might result in a decrease in cold-weather mortality (Medina-Ramón and Schwartz, 2007; Patz et al., 2000) but evidence is lacking, and McGeehin and Mirabelli (2001) note that "the relationship between winter weather and mortality is difficult to interpret." A lack of adequate heating is a concern if extreme weather events lead to blackouts or if economic strains make fuel poverty[8] more common. This may be a particular issue for elderly populations because physiological changes associated with the aging process make them more vulnerable to the effects of cold (Press, 2003).

Evidence indicates that cold weather is associated with an excess of mortality (Analitis et al., 2008; Anderson and Bell, 2009; Donaldson and Keatinge, 1997; Huynen et al., 2001; Kloner et al., 1999). Potential causes, in cases where hypothermia can be ruled out, include cardiovascular death due to higher blood pressures resulting from lower core body temperatures (Barnett, 2007; Barnett et al., 2005; Danet et al., 1999; Donaldson et al., 1997; Medina-Ramón and Schwartz, 2007; Press, 2003). An increase of plasma fibrinogen during the winter has also been found to increase instance of ischemic heart disease (Woodhouse et al., 1994). And O'Neil and colleagues (2003) found an association between cold temperatures and respiratory-disease mortality in a hierarchical model that factored geographic location and socioeconomic variables.

Cold weather is not anticipated to be a climate change issue and cold weather exposures are not further explored in this chapter. However, two other chapters of this report address issues indirectly related to climate change, cold-weather conditions and health: Chapter 4 discusses adverse exposures associated with extreme weather events, including the use of unvented space heaters, back-up electrical power generators, and biofuel stoves indoors, and Chapter 6 talks about the influence of seasonality on the availability and spread of infectious agents.

CLIMATE-CHANGE ADAPTATION AND MITIGATION MEASURES

Protection from the adverse effects of heat exposure requires the ability to lower core temperature and often involves maintaining or moving to a temperate space. Many cities, for example, have heat-emergency plans that include cooling centers where people can seek shelter. Approaches for creating or maintaining a safe thermal environment are outlined below.

Heating, Ventilating, and Air-Conditioning Approaches

Demonstration projects and research suggest that innovations in the design of mechanical systems and buildings may yield reduced HVAC-system energy use while enhancing occupant comfort, health, and productivity. They include both mature and newly developed technologies:

[8] Fuel poverty is defined as spending more than 10% of income on heating a home to an adequate level of warmth (Press, 2003)

- Mixed-mode or hybrid mechanical systems that support natural ventilation (Axley, 2001; WHO, 2009).
- Economizer-cycle HVAC (Fisk and Seppänen, 2007).
- Water-based cooling systems, including fan-coil, radiant, and induction systems (Costelloe and Finn, 2003).
- High efficiency, low-pressure–drop filtration (Fisk, 2009).
- Displacement ventilation (Schiavon, 2009).
- Passive stack and solar chimney systems (Russell et al., 2005).
- Geothermal heat exchangers (Eicker and Vorschulze, 2009).
- Earth-tube exchangers (Darkwa et al., 2011; Zmeureanu and Wu, 2007).

Mudarri's Environmental Protection Agency white paper (2010) notes that HVAC approaches like those vary in their ease of implementation: some constitute straightforward upgrades of existing systems, and others can be achieved only through building renovation or are feasible only for new construction. The cost effectiveness of the measures is strongly linked to the price of energy.

Building-Design and Setting Approaches

Architects, builders, and city planners have several tools at their disposal for influencing the amount of heat absorbed by buildings and the amount dissipated by them. Some are ancient and of established efficacy. Traditional construction in warm climates—including the American Southwest, southern Europe, and the Middle East—has long used light, reflective colors for exteriors. Synnefa et al. (2007) estimated that increasing roof reflectivity from its current 10–20% to 60% through the use of cool-colored materials and coatings could reduce cooling-energy use by more than 20%. Models developed by Akbari and colleagues (2001) suggest that introducing additional trees and reflective or light-colored building and road surfaces to urban environments would not only lower energy use but would lessen heat-island effect. Installation of green roofs composed of soil substrate and plants (Oberndorfer, et al. 2007) and regionally and seasonally appropriate use of landscape elements and trees to block summer sunlight but permit winter solar heating have also been shown to reduce cooling and heating loads and peak energy demands (Akbari, 2002) and to lower concentrations of air pollutants (Nowak et al., 2006; Yang et al., 2008).

Building-performance simulation (BPS) tools constitute another approach to managing heat through passive, low-energy means. BPS models estimate energy and mass flows in buildings as functions of the characteristics of a building and the space around it. Reinhart et al. (2010), in a presentation before the committee, noted the utility of such simulations in understanding how neighboring buildings may affect heating loads and local wind patterns and thus influence whether natural ventilation can be used successfully. Climate-change projections can be married to BPS models to estimate the benefits of particular building or site modifications in mitigating the effects of climate change.

Passive Survivability

Passive survivability is a term coined by Alex Wilson (2005) to describe "a building's ability to maintain critical life-support conditions in the event of extended loss of power, heating fuel, or water, or in the event of extraordinary heat spells." Interest in the concept may have been stimulated in part by reports of deaths in sealed buildings that were left without power in the wake of Hurricane Katrina. The elements of passive survivability include provision for natural ventilation even if a building was designed to operate with a mechanical HVAC system; resilience in the face of extreme weather; high levels of insulation and other high-performance building-envelope features; minimization of cooling loads through building geometry, landscaping, and thermal mass; passive solar heating; and natural daylight (GSA, 2010; Wilson, 2006). Santamouris et al. (2007) note that such features are especially important in low-income housing, where residents are more likely to suffer from heat stress and poor indoor environmental quality.

Passive survivability has gained currency in the General Services Administration (GSA), which manages buildings for the federal government. Testimony from its administrator in 2007 indicated that GSA was undertaking initiatives to address facility passive survivability (Doan, 2007), and its 2010 *Facilities Standards for the Public Buildings Service* identified it as a best-practice strategy (GSA, 2010).

Synthesis

A number of techniques for reducing the health risks and productivity costs associated with uncomfortable or unsafe indoor thermal environments are available. They include both well-established low-technology passive strategies and cutting-edge design and technology innovations. Many of the approaches identified above yield additional benefits, including lower energy use and costs (with concomitantly reduced generation of greenhouse gases) and better building ventilation, which is associated with lower incidence of respiratory and other health problems. The best passive approaches for a given building will depend on its age, location, and use and on the resources available to implement changes.

Warmer outdoor conditions and more frequent and severe weather events will stimulate greater interest in using those techniques to mitigate effects or adapt to changing conditions. Climate change may also affect the economics of implementation as the price of energy increases and as the human and social costs of inaction become untenable.

CONCLUSIONS

On the basis of its review of the papers, reports, and other information presented in this chapter, the committee has reached the following conclusions regarding the health effects of alterations in indoor environmental quality due to thermal conditions:

- Thermal stress has well-documented adverse health effects, and is responsible for excess mortality among exposed persons.
- Health, economic, and social factors make certain populations particularly vulnerable to exposure to temperature extremes and to the adverse consequences of such exposure, and may limit their ability to mitigate or seek shelter from health-threatening conditions. The elderly, those in poor health, and the poor are especially at risk. Those populations experience temperature extremes almost exclusively in indoor environments.
- Air conditioning provides protection from the heat, and some types also offer protection from high concentrations of outdoor pollutants. However air conditioning is associated with higher reported prevalences of some ailments, perhaps because of contaminants in HVAC systems. No general conclusion can thus be drawn about the effect of air conditioning on adverse biologic or chemical exposures indoors.
- Little research has addressed the effects of climate change on building energy use and occupant health. Available information indicates that changing conditions may have the following effects:
    - Buildings that are currently ventilated naturally will need to use some form of air conditioning.
    - Buildings that have air conditioning will need to use it more often, reducing natural ventilation.
    - People in buildings that do not have air conditioning will be exposed to extreme heat conditions more often.
- Many buildings in warm zones of the United States already have air conditioning. However, there is concern that peak energy demands during extreme heat events and an increased frequency of extreme weather events may result in more frequent power outages that expose large numbers of persons to potentially dangerous conditions indoors.

- Temperate indoor conditions (70–72°F or 21–22°C) are associated with higher office and school productivity than colder or warmer environments.
- Several technologies and building-design and -siting approaches can provide control of the indoor environment with lower energy costs and greater health benefits than systems typically in use today. No approach will work in all circumstances; the best strategies will depend on building use and on local and occupant circumstances.
- No matter which approach is used to maintain safe indoor environmental conditions, it is important to ensure that the conditions are sustained when failures in building systems or power outages disable mechanical ventilation—something that may happen more often if climate change leads to more instances of extreme weather conditions or unsustainable loads on the electric grid due to extreme outdoor temperatures

REFERENCES

Akbari H, Pomerantz M, Taha H. 2001. Cool surfaces and shade trees to reduce energy use and improve air quality in urban areas. *Solar Energy* 70(3): 295-310.

Akbari H. 2002. Shade trees reduce building energy use and CO2 emissions from power plants. *Environmental Pollution* 116(1S):S119-S126.

Aldous MB, Holberg CJ, Wright AL, Martinez FD, Taussig LM, Group Health Medical Associates. 1996. Evaporative cooling and other home factors and lower respiratory tract illness during the first year of life. *American Journal of Epidemiology* 143(5):423-430.

Analitis A, Katsouyanni K, Biggeri A, Baccini M, Forsberg B, Bisanti L, Kirchmayer U, Ballester F, Cadum E, Goodman PG, Hois A, Sunyer J, Tjittanen P, Michelozzi P. 2008. Effects of cold weather on mortality: Results from 15 European cities within the PHEWE project. *American Journal of Epidemiology* 168(12):1397-408.

Anderson BG, Bell ML. 2009. Weather-related mortality: How heat, cold, and heat waves affect mortality in the United States. *Epidemiology* 20(2):205-213.

ASHRAE (American Society of Heating, Refrigerating and Air-Conditioning Engineers). 2004. *ANSI/ASHRA Standard 55-2004: Thermal environmental conditions for human occupancy.* Atlanta, GA: ASHRAE.

Axley JW. 2001. *Application of natural ventilation for U.S. commercial buildings—Climate suitability design strategies & methods modeling studies.* Gaithersburg, MD: National Institute of Standards and Technology.

Barnett AG, Dobson AJ, McElduff P, Salomaa V, Kuulasmaa K, Sans S. 2005. Cold periods and coronary events: An analysis of populations worldwide. *Journal of Epidemiology & Community Health* 59(7):551-557.

Barnett AG. 2007. Temperature and cardiovascular deaths in the US elderly: Changes over time. *Epidemiology* 18(3):369-372.

Basu R, Samet JM. 2002. Relation between elevated ambient temperature and mortality: A review of the epidemiologic evidence. *Epidemiologic Reviews* 24(2):190-202.

Basu R, Ostro BD. 2008. A multicounty analysis identifying the populations vulnerable to mortality associated with high ambient temperature in California. *American Journal of Epidemiology* 168(6):632-637.

Bell ML, Dominici F. 2008. Effect modification by community characteristics on the short-term effects of ozone exposure and mortality in 98 US communities. *American Journal of Epidemiology* 167(8):986-997.

Bell ML, Ebisu K, Peng RD, Dominici F. 2009. Adverse health effects of particulate air pollution modification by air conditioning. *Epidemiology* 20(5):682-686.

Brown, RE, Koomey JG. 2003. Electricity use in California: Past trends and present usage patterns. *Energy Policy* 31(9):849-864.

Carberry PA, Shepherd AM, Johnson JM. 1992. Resting and maximal forearm skin of blood flow are reduced in hypertension. *Hypertension* 20:349-355.

Carmichael J, Shankel SW. 1985. Effects of non-steroidal anti-inflammatory drugs on prostaglandins and renal function. *The American Journal of Medicine* (6 Pt 1):992-1000.

CDC (Centers for Disease Control and Prevention). 2010. *Heart disease.* http://www.cdc.gov/nchs/fastats/heart.htm (accessed January 6, 2011).

Changnon SA, Kunkel KE, Reinke BC. 1996. Impacts and responses to the 1995 heat wave: A call to action. *Bulletin of the American Meteorological Society* 77:1497-1506.

Costelloe B, Finn D. 2003. Indirect evaporative cooling potential in air–water systems in temperate climates. *Energy and Buildings* 35(6):573-591.

Curriero FC, Heiner KS, Samet JS, Zeger S, Patz JA. 2002. Temperature and mortality in eleven cities of the eastern United States. *American Journal of Epidemiology* 155:80-87.

Danet S, Richard F, Montaye M, Beauchant S, Lemaire B, Grauz C, Cottel D, Marécaux N, Amouyel P. 1999. Unhealthy effects of atmospheric temperature and pressure on the occurrence of myocardial infarction and coronary deaths. A 10-year survey: The Lille-World Health Organization MONICA project (Monitoring trends and determinants in cardiovascular disease). *Circulation* 100(1):E1-E7.

Darkwa J, Kokogiannakis G, Magadzire CL, Yuan K. 2011. Theoretical and practical evaluation of an earth-tube (E-tube) ventilation system. *Energy and Buildings* 43(2-3):728-736.

de Dear R, Brager GS. 1998. Developing an adaptive model of thermal comfort and preference. *ASHRAE Transactions* 104(1):145-167.

Dematte JE, O'Mara K, Buescher J, Whitmey CG, Forsythe S, McNamee T, Adiga RB, Ndukwu IM. 1998. Near-fatal heat stroke during the 1995 heat wave in Chicago. *Annals of Internal Medicine* 129(3):173-181.

Doan L. 2007. *Administration's response to climate change and energy independence.* Statement of Lurita Doan, Administrator, U.S. General Services Administration, Before the Committee on Transportation and Infrastructure, U.S. House of Representatives, May 11, 2007. http://www.gsa.gov/portal/content/102626 (accessed February 24, 2011).

DOE (Department of Energy). 2000. *Trends in residential air-conditioning usage from 1978 to 1997.* http://www.eia.doe.gov/emeu/consumptionbriefs/recs/actrends/recs_ac_trends.html (accessed February 8, 2011).

DOE. 2008. *2005 Residential energy consumption survey Table HC2.7 Air conditioning usage indicators by type of housing unit, 2005.* http://www.eia.doe.gov/emeu/recs/recs2005/hc2005_tables/hc7airconditioningindicators/excel/tablehc2.7.xls (accessed February 8, 2011).

Donaldson G, Robinson D, Allaway S. 1997. An analysis of arterial disease mortality and BUPA health screening data in men, in relation to outdoor temperature. *Clinical Science* 92:261-268.

Donaldson GC, Keatinge WR. 1997. Early increases in ischaemic heart disease mortality dissociated from and later changes associated with respiratory mortality after cold weather in south east England. *Journal of Epidemiology & Community Health* 51(6):643-648.

ED (Department of Education). 2007. *Public school principals report on their school facilities: Fall 2005 statistical analysis report.* National Center for Education Statistics. http://nces.ed.gov/pubs2007/2007007.pdf (accessed February 9, 2011).

Eggers FJ, Thackeray A. 2007. *32 Years of housing data.* Report prepared for US Department of Housing and Urban Development, Office of Policy Development and Research. Bethesda, MD: Econometrica, Inc.

EIA (US Energy Information Administration) 2007. *Commercial buildings energy consumption survey.* Washington, DC: EIA.

EIA. 2010a. Air conditioning characteristics by climate-zone, 2005. http://www.eia.gov/emeu/recs/recs2005/hc2005_tables/hc6airconditioningchar/pdf/tablehc9.6.pdf (accessed July 18, 2011).

EIA. 2010b. Air conditioning characteristics by type of housing unit, 2005. http://www.eia.gov/emeu/recs/recs2005/hc2005_tables/hc6airconditioningchar/pdf/alltables.pdf (accessed July 18, 2011).

Eicker U, Vorschultze C. 2009. Potential of geothermal heat exchangers for office building climatisation. *Renewable Energy* 34:1126-1133.

EPA (US Environmental Protection Agency). 2009a. *Heat island effect: Glossary.* http://www.epa.gov/hiri/resources/glossary.htm (accessed February 24, 2011).

EPA. 2009b. *Indoor air quality tools for schools reference guide.* Washington, DC: EPA.

Fealy RD, Low PA, Thomas JE. 1989. Thermoregulatory sweating abnormalities in diabetes mellitus. *Mayo Clinic Proceedings* 64:617-628.

Fisk WJ, Seppänen OA. 2007. *Providing better indoor environmental quality brings economic benefits.* Published in *Proceedings of Climate 2007. Well Being Indoors*, June 10–14, 2007, Helsinki. Paper A01. http://eetd.lbl.gov/ied/sfrb/pdfs/performance-1.pdf (accessed February 9, 2011).

Fisk WJ. 2009. *Climate change, energy efficiency, and IEQ: Challenges and opportunities for ASHRAE.* Berkeley, CA: Lawrence Berkeley National Laboratory.

Fouillet A, Rey G, Laurent F, Pavillon G, Bellec S, Guihenneue-Jouyaux C, Clavel J, Jougla E, Hémon D. 2006. Excess mortality related to the August 2003 heat wave in France. *International Archives of Occupational and Environmental Health* 80:16-24.

Fryar CD, Hirsch R, Eberhardt MS, Yoon SS, Wright JD. 2010. Hypertension, high serum total cholesterol, and diabetes: Racial and ethnic prevalence differences in U.S. adults, 1999-2006. *NCHS Data Brief* (36):1-8.

GSA (US General Services Administration). 2010. *Facilities standards for the public service buildings P100.* Washington, DC: GSA.

Hajat S, Kovats RS, Lachowycz K. 2007. Heat-related and cold-related deaths in England and Wales: Who is at risk? *Occupational and Environmental Medicine* 64(2):93-100.

Hoffmann B, Hertel S, Boes T, Weiland D, Jockel, KH. 2008. Increased cause-specific mortality associated with 2003 heat wave in Essen, Germany. *Journal of Toxicology and Environmental Health A* 71:759-765.

Hummelgaard J, Juhl P, Sæbjörnsson KO, Clausen G, Toftum J, Langkilde G. 2007. Indoor air quality and occupant satisfaction in five mechanically and four naturally ventilated open-plan office buildings. *Building and Environment* 42(12):4051-4058.

Huynen MM, Martens P, Schram D, Weijenberg MP, Kunst AE. 2001. The impact of heat waves and cold spells on mortality rates in the Dutch population. *Environmental Health Perspectives* 109(5):463-470.

IPCC (International Panel on Climate Change). 2007. Chapter 9. Understanding and attributing climate change. In *Climate change 2007—The physical science basis.* Contribution of Working Group I to the Fourth Assessment Report of the Intergovernmental Panel on Climate Change, edited by Solomon S, Qin D, Manning M, Chen Z, Marquis M, Averyt KB, Tignor M, Miller HL. Cambridge, United Kingdom and New York, NY, USA: Cambridge University Press. http://www.ipcc.ch/pdf/assessment-report/ar4/wg1/ar4-wg1-chapter9.pdf (accessed February 9, 2011).

Kenney WL, Munce TA. 2003. Invited review: Aging and human temperature regulation. *Journal of Applied Physiology* 95(6):2598-2603.

Kenny GP, Yardley J, Brown C, Sigal R, Jay O. 2010. Heat stress in older individuals and patients with common chronic diseases. *CMAJ: Canadian Medical Association Journal* 182(10):1053-1060.

Khalaj B, Lloyd G, Sheppeard V, Dear K. 2010. The health impacts of heat waves in five regions of New South Wales, Australia: A case-only analysis. *International Archives of Occupational and Environmental Health* 83(7):833-842.

Kilbourne EM, Choi K, Jones S, Thacker SB. 1982. Risk factors for heatstroke. A case-control study. *Journal of the American Medical Association* 247(24):3332-3336.

Klinenberg E. 2002. *Heat wave: A social autopsy of disaster in Chicago (Illinois)*. Chicago: The University of Chicago Press.

Kloner RA, Poole WK, Perritt RL. 1999. When throughout the year is coronary death most likely to occur? A 12-year population-based analysis of more than 220 000 cases. *Circulation* 100(15):1630-1634.

Knowlton K, Rotkin-Ellman M, King G, Margolis HG, Smith D, Solomon G, Trent R, English P. 2009. The 2006 California heat wave: Impacts on hospitalizations and emergency department visits. *Environmental Health Perspectives* 117(1):61-67.

Kovats RS, Hajat S. 2008. Heat stress and public health: A critical review. *Annual Review of Public Health* 29:41-55.

Lan L, Wargocki P, Lain Z. 2010. Quantitative measurement of productivity loss due to thermal discomfort. *Energy and Buildings* 43(5):1057-1062.

Lopez R., Goldoftas, B. 2009. The urban elderly in the United States: Health status and the environment. *Reviews on Environmental Health* 24:47-57.

Luber G, McGeehin M. 2009. Climate change and extreme heat events. *American Journal of Preventative Medicine* 35(5):429-435.

Martinez BF, Annest JL, Kilbourne EM, Kirk ML, Lui K-J, Smith ZM. 1989. Geographic distribution of heat-related deaths among elderly persons. Use of county-level dot maps for injury surveillance and epidemiologic research. *Journal of the American Medical Association* 262(16):2246-2250.

McGeehin MA, Mirabelli M. 2001. The potential impacts of climate variability and change on temperature-related morbidity and mortality in the United States. *Environmental Health Perspectives* 109(2):185-189.

Medina-Ramón M, Schwartz J. 2007. Temperature, temperature extremes, and mortality: A study of acclimatization and effect modification in 50 United States cities. *Occupational & Environmental Medicine* 67:827-833.

Mendell MJ, Lei-Gomez Q, Mirer AG, Seppänen O, Brunner G. 2008. Risk factors in heating, ventilating, and AC systems for occupant symptoms. *Indoor Air* 18:301-316.

Metzger KB, Ito K, Matte TD. 2010. Summer heat and mortality in New York City: How hot is too hot? *Environmental Health Perspectives* 118(1):80-86. Comment in 118(1):A35.

MMWR (Morbidity and Mortality Weekly Report). 1998. Community needs assessment and morbidity surveillance following an ice storm—Maine, January 1998. *MMWR* 47:351-354.

Morello-Frosch R, Jesdale B. 2008. Unpublished impervious surface and tree cover data. Data for this analysis were derived from: US Geological Survey's National Land Cover Dataset 2001. http://www.mrlc.gov/nlcd.php (accessed June 20, 2007); and ESRI's ArcMap census boundary files http://www.census.gov/geo/www/cob/bdy_files.html (accessed June 6, 2008).

Mudarri D. 2010. *Public health consequences and cost of climate change impacts on indoor environments*. Washington, DC: US Environmental Protection Agency.

Naughton MP, Henderson A, Mirabelli MC, Kaiser R, Wilhelm JL, Kieszak SM, et al. 2002. Heat-related mortality during a 1999 heat wave in Chicago. *American Journal of Preventative Medicine* 22(4):221-227.

Nicol JF, Humphreys MA. 2002. Adaptive thermal comfort and sustainable thermal standards for buildings. *Energy and Buildings* 34(6):563-572.

Nowak D, Crane D, Stevens J. 2006. Air pollution removal by urban trees and shrubs in the United States. *Urban Forestry and Urban Greening* 4:115-123.

NRC (National Research Council). 2010. *America's climate choices: Advancing the science of climate change.* Washington, DC: The National Academies Press.

NWS (National Weather Service). 2010. *NOAA's National Weather Service heat index.* http://www.nws.noaa.gov/om/heat/heatindex.shtml (accessed January 25, 2011).

Oberndorfer E, Lundholm J, Bass B, Coffman RR, Doshi H, Dunnett N, Gaffin S, Kohler M, Liu KKY, Rowe B. 2007. Green roofs as urban ecosystems: Ecological structures, functions, and services. *BioScience* 57(10):823-833.

O'Neill MS, Zanobetti A, Schwartz J. 2003. Modifiers of the temperature and mortality association in seven US cities. *American Journal of Epidemiology* 157:1074-1082.

Patz JA, McGeehin MA, Bernard SM, Ebi KL, Epstein PR, Grambsch A, Gubler DJ, Reiter P, Romieu I, Rose JB, Samet JM, Trtanj J. 2000. The potential health impacts of climate variability and change for the United States: Executive summary of the report of the Health Sector of the U.S. National Assessment. *Environmental Health Perspectives* 108(4):367-376.

Pearlman RA, Uhlmann RF. 1988. Quality of life in chronic diseases: Perceptions of elderly patients. *Journal of Gerontology* 43(2):M25-M30.

Petrofsky JS, Lee S, Patterson C, Cole M, Stewart B. 2005. Sweat production during global heating and during isometric exercise in people with diabetes. *Medical Science Monitor* 11:CR515-CR521.

Phelan, JC, Link BG, Diez-Roux A, Kawachi I, Levin B. 2004. Fundamental causes of social inequalities in mortality: A test of the theory. *Journal of Health and Social Behavior* 45(3):265-285.

Press V. 2003. *Fuel poverty and health.* London, UK: National Heart Forum.

Preziosi P, Czernichow S, Gehanno P, Hercberg S. 2004. Workplace air-conditioning and health services attendance among French middle-aged women: A prospective cohort study. *International Journal of Epidemiology* 33:1120-1123.

Reid CE, O'Neill MS, Gronlund CJ, Brines SJ, Brown DG, Diez-Roux AV, Schwartz J. 2009. Mapping community determinants of heat vulnerability. *Environmental Health Perspectives* 177(11):1730-1736.

Reinhart C, Holmes S, Park C. 2010. *Climate change & (solar) architecture.* Presentation before the Committee on the Effect of Climate Change on Indoor Air Quality and Public Health on June 7, 2010.

Ruddell DM, Harlan SL, Grossman-Clarke S, Buyantuyev A. 2010. Risk and exposure to extreme heat in microclimates of Phoenix, AZ. In *Geospatial contributions to urban hazard and disaster analysis*, edited by Showalter PS, Lu Y. London, NY: Springer Dordrecht Heidelberg.

Russell M, Sherman M, Rudd A. 2005. *Review of residential ventilation technologies.* Berkeley, CA: Ernest Orlando Lawrence Berkeley National Laboratory.

Sahakian N, Park J, Cox-Ganser J. 2009. Respiratory morbidity and medical visits associated with dampness and air-conditioning in offices and homes. *Indoor Air* 19(1):58-67.

Santamouris M, Pavloua K, Synnefaa A, Niachoua K, Kolokotsab D. 2007 Recent progress on passive cooling techniques—Advanced technological developments to improve survivability levels in low-income households. *Energy and Buildings* 39(Special Issue S1):859-866.

Schiavon S. 2009. *Energy saving with personalized ventilation and cooling fan.* Doctoral dissertation, Padua: University of Padua Department of Applied Physics.

Schuman SH. 1967. Patterns of urban heat wave deaths and implications for prevention: Data from New York and St Louis during July, 1966. *Environmental Research* 5:59-75.

Scott MJ, Huang YJ. 2007. Effects of climate change on energy use in the United States. In *Effects of climate change on energy production and use in the United States,* edited by Wilbanks TJ, Bhatt V, Bilello DE, Bull SR, Ekmann J, Horak WC, Huang YJ, Levine MD, Sale MJ, Schmalzer DK, Scott MJ. Synthesis and Assessment Product 4.5. U.S. Climate Change Science Program, Washington, DC, pp. 8-44.

Semenza J. 1999. Excess hospital admissions during the July 1995 heat wave in Chicago. *American Journal of Preventative Medicine* 16(4):269-277.

Semenza JC, Rubin CH, Falter KH, Selanikio JD, Flanders WD, Howe HL, Wilhelm JL. 1996. Heat-related deaths during the July 1995 heat wave in Chicago. *New England Journal of Medicine* 335(2):84-90.

Shonkoff SB, Morello-Frosch R, Pastor M, Sadd J. 2009. *Draft Paper: Environmental health and equity impacts from climate change and mitigation policies in California: A review of the literature.* California Climate Change Center.

Stansberry KB, Hill MA, Shapiro SA, McNitt PM, Bhatt BA, Vinik AI. 1997. Impairment of peripheral blood flow responses in diabetes resembles an enhanced aging effect. *Diabetes Care* 20:1711-1716.

Steadman RG. 1979. The assessment of sultriness. Part I: A temperature-humidity index based on human physiology and clothing science. *Journal of Applied Meteorology* 18:861-873.

Synnefa A, Santamouris M, Akbari H. 2007. Estimating the effect of using cool coatings on energy loads and thermal comfort in residential buildings in various climatic conditions. *Energy and Buildings* 39(11):1167-1174.

Tan J, Zheng Y, Song G, Kalkstein LS, Kalkstein AJ, Tang X. 2007. Heat wave impacts on mortality in Shanghai, 1998 and 2003. *International Journal of Biometeorology* 51(3):193-200.

USCB (US Census Bureau). 2004. *Housing vacancies and home ownership (CPS/HVS).* http://www.census.gov/hhes/www/housing/hvs/annual97/ann97def.html (accessed February 24, 2011).

USCB. 2009. *American Housing Survey (AHS).* http://www.census.gov/hhes/www/housing/ahs/ahs09/ahs09.html (accessed February 24, 2011).

USGCRP (US Global Change Research Program). 2009. *Global climate change impacts in the United States.* New York: Cambridge University Press.

Vandentorren S, Bretin P, Zeghnoun A, Mandereau-Bruno L, Croisier A, Cochet C, Ribéron J, Siberan I, Declercq B, Ledrans M. 2006. August 2003 heat wave in France: Risk factors for death of elderly people living at home. *European Journal of Public Health* 16:583-591.

Verdú E, Ceballos D, Vilches JJ, Navarro X. 2000. Influence of aging on peripheral nerve function and regeneration. *Journal of the Peripheral Nervous System* 5(4):191-208.

WHO (World Health Organization). 2009. *Natural ventilation for infection control in healthcare settings.* Geneva: WHO Press.

Wilson A. 2005. Passive survivability. *Environmental Building News.* December 1. http://www.buildinggreen.com/auth/article.cfm/2005/12/1/Passive-Survivability/ (accessed February 23, 2011).

Wilson A. 2006. Passive survivability: A new design criterion for buildings. *Environmental Building News.* May 1. http://www.buildinggreen.com/auth/article.cfm/2006/5/3/Passive-Survivability-A-New-Design-Criterion-for-Buildings/ (accessed February 23, 2011).

Woodhouse PR, Khaw KT, Plummer M, Foley A, Meade TW. 1994. Seasonal variations of plasma fibrinogen and factor VII activity in the elderly: Winter infections and death from cardiovascular disease. *Lancet* 343(8895):435-439.

Yang J, Yu Q, Gong P. 2008. Quantifying air pollution removal by green roofs in Chicago. *Atmospheric Environment* 42(31):7266-7273.

Zmeureanu R, Wu X. 2007. Energy and exergy performance of residential heating systems with separate mechanical ventilation. *Energy* 32:187-195.

8

Building Ventilation, Weatherization, and Energy Use

High energy costs and climate-change mitigation efforts are creating pressures to decrease ventilation rates in buildings as a means of reducing the energy used to cool or warm indoor air. This chapter concentrates on the interrelated issues of building energy use, emissions from building materials, weatherization, and ventilation and on how they affect occupants. It addresses energy consumption in buildings, the means used to tighten buildings, programs to enhance the energy efficiency of buildings and reduce harmful emissions from building components, the training of personnel who implement weatherization programs, and the effect of tightening on ventilation, indoor environmental quality, and occupant health and productivity. The chapter concludes with the committee's observations regarding those issues.

Ventilation affects indoor levels of air pollutants, indoor moisture levels, exposures to biologic agents, and the thermal environment of homes. Research on those topics as opposed to ventilation itself is addressed in Chapters 4–7.

ENERGY USE IN BUILDINGS

Energy use in buildings has been a concern in the United States since the oil embargoes of the 1970s but has gained new currency in recent years as a result of rising costs and an interest in limiting greenhouse-gas emissions. The Department of Energy (DOE) tracks trends in energy use. Its 2009 *Buildings Energy Data Book*, which has data through 2006, notes that the dominant uses vary between residential and commercial structures[1] (DOE, 2009). As noted in Table 8-1, the dominant uses of energy in the residential sector are ambient space heating (about 26%) and cooling, water heating, and lighting (each about 12–13%). In commercial buildings, lighting is the dominant category at about 25%, but space heating, cooling, and mechanical ventilation together account for more than 31%. DOE also estimates emissions of carbon dioxide (CO_2), a greenhouse gas, from burning fossil fuels to generate energy (mainly natural gas on site and natural gas and coal for electricity production). Those figures are listed in Table 8-1, and they track the energy-use numbers closely. All told, building CO_2 emissions in 2006 accounted for 38% of total US CO_2 emissions—20% contributed by residential buildings, 18% by commercial structures.

BUILDING WEATHERIZATION

Weatherization describes the steps taken during building design or retrofit to increase energy efficiency by limiting unintended air and heat exchange between the indoor and outdoor environments. Because those steps generally entail closing gaps in the building envelope, the process is also referred to as tightening. This section describes some of the means typically used to tighten buildings and the effect of tightening on ventilation.

Strategies for Tightening Buildings

There are four common methods for reducing unplanned air leakage in buildings.

Air-tighten the enclosure. Sealing cracks, gaps, and holes in the building envelope with vapor barriers, and other construction changes reduce the amount of air that accidentally leaks in or out. In many US climates, this saves substantial amounts of energy. Sherman and McWilliams (2007) determined that around one-third of the energy used for heating and cooling is due to accidental air leakage. There are far fewer measurement data on accidental air leakage in commercial buildings, but it is reported to be around 20–30% (range, 0–58%) of the heating or cooling energy used (Edwards and Hamilton, 1993; Emmerich, 2005; Shaw, 1995). In a study of several California buildings, Mowris and Fisk (1988) observed that accidental air leakage made up 0–30% of the total air-exchange rate. Persily and Norford (1987) found leakage of 31–58% in a three-story office building. About 20–40% of the air leakage can be sealed in existing residential

[1] There is, of course, great variation among buildings in these general categories; building age, material, size, location, and predominant use are important factors.

TABLE 8-1 Percentage of Total Energy Use and Carbon Dioxide Emissions Attributable to Specific Applications in US Buildings in 2006 (DOE, 2010)

	Energy Use			Carbon Dioxide Emissions		
	All	Residential	Commercial	All	Residential	Commercial
Space heating	19.8	26.4	12.1	18.8	24.6	12.2
Lighting	17.7	11.6	24.8	18.1	12.0	25.2
Space cooling	12.7	13.0	12.6	13.0	13.4	12.5
Water heating	9.6	12.5	6.3	9.4	12.4	6.0
Electronics	7.8	8.1	7.5	8.0	8.4	7.6
Refrigeration	5.8	7.2	4.1	5.9	7.4	4.2
Cooking	3.4	4.7	2.0	3.4	4.7	1.9
Wet cleaning[a]	3.3	6.2	—	3.4	6.4	—
Mechanical ventilation	2.8	—	6.7	2.9	—	6.2
Computers	2.3	1.0	3.8	2.4	1.0	3.9
Other	8.5	3.6	13.2	8.4	3.8	12.6
Attributable to buildings but not directly to specific end uses	6.3	5.7	6.9	6.4	5.9	7.9

[a] Primarily automatic washers, dryers, and dishwashers.

and commercial buildings; in new construction, it is feasible to seal about 90% of potential leakage in typical stock (Spengler, 2010).

Seal air-distribution systems. Holes and gaps in air handlers, supply and return ducts, and plenums[2] lead to leakage in buildings. If the air-handling system is off, then they behave like any other leak. When an air handler is on, leaks are exacerbated by the greater pressure difference across holes or gaps. Cummings et al. (1996) reported that measured duct leakage in commercial buildings averaged about 80 ft^3/min at 25 pascals/100 ft^2 of duct surface area; the largest outdoor-air infiltration rates were in vented spaces, such as attics, crawlspaces, mechanical closets, and wall cavities. Indoor relative humidity may increase if hot and humid outdoor air infiltrates these spaces. In response, air-conditioning units may have longer run times to correct the imbalance and thus waste energy.

Manage indoor–outdoor air-pressure differences. If airflow through heating, ventilating, and air-conditioning (HVAC) equipment results in excessively pressurized or depressurized zones, rooms, or building cavities, then indoor air may be forced out or drawn in through the building enclosure. The most common example is return air plenums. Any air leak in exterior walls that bound a return plenum becomes an accidental outdoor-air intake when the air handler is operating. If a mechanical room is used as a mixing chamber for return and outdoor air, the room is likely to be depressurized by 10–30 pascals and may be depressurized by as much as 90 pascals (Spengler, 2010). Another example common in residential and small commercial buildings is a duct layout that includes supply diffusers in every room and air returns in corridors. When doors to the corridor are closed, the rooms are pressurized, and the corridors are depressurized. The combination of that dynamic and duct leakage to the outside can greatly increase the air-exchange rate in a building. In a study of unplanned airflows in 70 commercial buildings, 8 had air-exchange rates of 2 to 10 air changes per hour (ACH) when the air handlers were running (Cummings et al., 1996). Similar results have been reported for residential buildings. A research project on 91 Florida homes found that the average air-exchange rate went from 0.21 ACH with air handlers off to 0.91 ACH with air handlers on (Cummings and Tooley, 1989). Depending on the circumstances, poorly managed pressure differences may decrease desired circulation of outdoor air indoors or increase energy costs through excessive intrusion of outdoor air.

Replace atmospherically vented combustion equipment with high-efficiency combustion equipment in residential buildings. Atmospherically vented combustion equipment typically vents through a chimney. The

[2] A plenum is a space in which a building's supply or return air is mixed or moves; it can be a duct, a joist space, an attic, a crawlspace, or a wall cavity (EPA, 2011b).

chimney ventilates the equipment room and, when the equipment is not running, exhausts air. When the equipment is running, it ventilates at a much higher rate because of the high flue-gas temperature. Measurements made in the 1980s found typical flows of air through chimneys of 50–100 ft^3/min (Spengler, 2010). High-efficiency combustion equipment, such as condensing furnaces, does not have a chimney—it vents to the outdoors through pipes in a side wall. The combustion gases are vented to the outside through a small fan, which typically runs at about 25 ft^3/min. When such furnaces are not firing, the flows are essentially zero. That dynamic affects single-family buildings, some low-rise multifamily buildings, and small commercial buildings that are constructed with residential methods. If the equipment is in a basement or crawlspace in a climate that has a substantial heating season, ventilation through the chimney is often controlling humidity that enters through the foundation. That does not, however, apply to sealed combustion units (which draw no air from the mechanical space) and heat pumps (which need no vents, because there is no combustion).

Effects of Tightening on Ventilation

Lawrence Berkeley National Laboratory (LBNL) investigators compiled a dataset on blower-door tests used to assess air leakage from homes across the country (Chan et al., 2003). Figure 8-1 illustrates the results of the home air-tightness tests expressed as whole-house air exchange vs the year when a home was built. The solid line represents the smoothed fit through the data. It shows that the air-exchange rate—as extrapolated from a blower-door pressurization test of air leakage—has been decreasing in homes built over the past 40 years. The trend in tighter house construction coincides with a housing boom in the United States in warm-climate areas, such as Atlanta, Las Vegas, and Phoenix, where air conditioning often reduced the dependence on natural ventilation.

An evaluation of air-tightness measurements in 201 commercial and institutional buildings in the United States by Emmerich and Persily (2005) found that the structures were tighter than the overall average for residential buildings but leakier than new residential construction. The authors state that "unlike the residential air tightness data, the database of US commercial building air tightness shows no indication of a trend toward tightness for newer buildings" (Persily and Emmerich, 2009).

ENERGY-EFFICIENCY PROGRAMS FOR BUILDINGS

Several government and private initiatives are aimed at reducing energy use in residences and commercial structures. Depending on the program, they may include energy audits, general or building-specific recommenda-

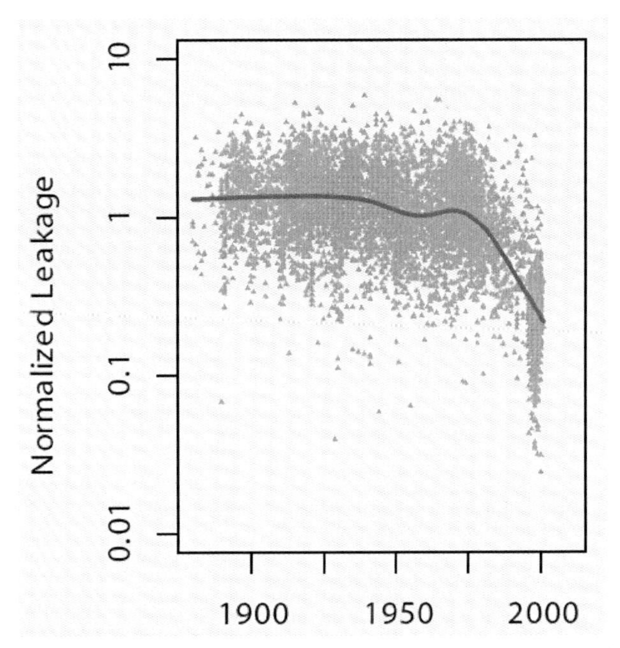

FIGURE 8-1 Normalized air leakage in a sample of homes (measured as air changes per hour) vs year when a home was built (Chan et al., 2003).

tions for action, and assistance in identifying or hiring contractors trained to perform remediations and upgrades. Improvements can include such weatherization measures as envelope and duct sealing, caulking, replacement of leaky windows, and increased insulation and such conservation steps as replacement of incandescent with compact fluorescent or LED lighting and appliance and HVAC upgrades. Information on some of the programs is summarized below.

National Weatherization Assistance Program

The national Weatherization Assistance Program (WAP) was launched in 1976 to help Americans with limited financial means to respond to rapidly increasing fuel prices during the oil embargoes of the 1970s. WAP weatherizes existing homes. Over the past 33 years, it has provided weatherization services to more than 6.4 million low-income households. Major funding comes from DOE, and additional support from a variety of sources, including the Low Income Home Energy Assistance Program block grants and energy utility programs (WAP, 2009b).

WAP conducts energy audits and selects appropriate energy-conserving measures that can be implemented for costs that do not exceed a capped dollar amount. In 2008, WAP provided services to around 100,000 households at a cap figure of $3,500 per unit. In 2009, that rose to 171,000 units. The target number of units weatherized for 2010 was 200,000 with a cap of $6,500 per unit (WAP, 2010).

WAP grantees and subgrantees use professionally trained staff and contractors. They make their own decisions on how training is provided. Although that creates some variability across the country, protocols for building assessment, weatherization measures, and quality-assurance procedures have evolved into a fairly consistent industry set of practices (WAP, 2009a). Training may be provided by independent weatherization trainers, inhouse technical trainers, or local or regional weatherization training centers. Typically, it is supplied by a mixture of the three. In 2009, 90% of states used state-agency staff for training, 75% used local-agency peers for training, and 70% used independent trainers (WAP, 2009a). Grantees also make their own decisions about certification. Some require certification by a national organization, such as the Residential Energy Services Network (RESNET) or the Building Performance Institute (BPI). Others provide their own certification, and 17 states do not require certification (WAP, 2009a).

Many of the training facilities for WAP also provide instruction to private-sector building-performance contractors. Twelve weatherization training centers in 11 states offer training that reaches beyond the WAP community. BPI supplies education through a network of training affiliate organizations, individual certifications, company accreditations, and quality-assurance programs. RESNET develops standards and certification for home-energy raters (WAP, 2009a). DOE is also planning to provide additional training for new workers in the weatherization field (WAP, 2009a).

ENERGY STAR

In 1992, the US Environmental Protection Agency (EPA) introduced ENERGY STAR as a voluntary labeling program designed to identify and promote energy-efficient products to reduce greenhouse-gas emissions. Computers and monitors were the first labeled products. However, in 1996, EPA partnered with DOE for some product categories, and the ENERGY STAR label was extended to new homes and commercial and industrial buildings (EPA, 2010b). In 1999, EPA, DOE, and the Department of Housing and Urban Development started Home Performance with ENERGY STAR, an energy-performance program for existing homes. The initiative provides guidelines and support for programs—often partnerships of government agencies, building-science technical-support organizations, building-performance contractors, and utility programs—to provide training and quality assurance for contractors who help homeowners to bring their properties up to ENERGY STAR standards for new construction. To receive an ENERGY STAR label, a home must be tested to demonstrate performance. More than 75,000 homes have been improved through the program (EPA, 2010b). More than a million US homes had received the ENERGY STAR label as of 2009, and 20% of the roughly 500,000 new homes constructed in 2009 were ENERGY STAR–labeled.

The ENERGY STAR Indoor Air Package is a label that adds items that address a broader array of indoor environmental quality issues—including moisture control, pest management, combustion safety, ventilation, emissions from building materials, and radon control—to the baseline ENERGY STAR program (EPA, 2010b). It was initiated in 2005 and intended to contribute to improved indoor air quality in new homes compared with code-built homes.

Other Programs

Nongovernment "green and affordable" housing programs are under way at the local and national levels. One example is the Enterprise Community Partnership, which serves low-income people in communities across the country and provides funds and expertise to enable developers to build and rehabilitate homes to be healthier, more energy-efficient, and consistent with sustainability criteria. The program started in 2004 and by 2009 had produced 17,500 new and renovated affordable homes (Enterprise Community Partners, 2010).

Many energy utility companies have energy-efficiency programs that aid residential, commercial, and institutional customers. The amount of money spent each year on such programs is large, totaling $5.3 billion in 2009 (Nevius et al., 2010). Program budgets vary widely by US Census region: states in the West account for 45% of the national total; in the Northeast, 25%; in the South, 17%; and in the Midwest, 13%. Utilities in California alone reported combined budgets of about $1.6 billion—30% of the national total. Programs were most likely to address energy-efficiency improvements in residential furnaces and boilers and in commercial and industrial lighting.

Commissioning and *retrocommisioning* are terms used to describe the usually independent evaluation of newly constructed or existing buildings (respectively) to determine whether they operate as designed or intended and whether they can be improved. It is more typically performed on commercial and public buildings. Examination of the energy efficiency of HVAC, lighting, plumbing, and other mechanical systems is a typical component of the process.

Effects of Programs on Energy Use and Employment

A 2010 LBNL case study of 14 energy-efficiency programs undertaken by state and local governments, utility companies, and nonprofit organizations found a wide range of participation and a maximum energy saving of about 15% for an individual home (Fuller et al., 2010). EPA reported that newly built homes implementing ENERGY STAR upgrades can realize up to 20% energy savings compared with conventional construction practice (EPA, 2007).

Effinger and Friedman (2010) summarized the findings of an LBNL report on retrocommissioning. The study—which comprised 112 buildings, including offices, hospitals, hotels, a retail space, and a school—found that measures to improve energy efficiency resulted in savings of 8–31% (median, 16%). The equipment affected included air-handling units, pumps, chillers, cooling towers, outside air-temperature sensors, and heat exchangers. The cost of the implemented measures and the retrocommissioning itself had a simple payback of 0.5–2.5 years (median, 1.1 years). The LBNL report itself (Mills, 2009) concluded that "these findings demonstrate that commissioning is arguably the single-most cost-effective strategy for reducing energy, costs, and greenhouse gas emissions in buildings today."

Residential energy upgrade programs through DOE, state energy offices, and mandated service of public utilities are expected to grow substantially over the next decade. California has committed to reducing energy use in existing homes by 40% by 2020 (CPUC, 2008). Nationally, jobs in the sector are expected to grow from around 114,000 person-years of employment (PYE) in 2008 to 200,000–380,000 PYE by 2020 (Goldman et al., 2010).

PRODUCT-LABELING AND BUILDING-CERTIFICATION PROGRAMS

Well before the green-building movement gained currency in the 1990s, indoor air quality concerns were recognized as more sealed buildings were constructed, ventilation rates were reduced, and new equipment, materials, coatings, and furnishings were introduced. In response to those concerns, governments and private organizations developed product-labeling, emission-testing, and building-certification systems to distinguish and market[3] healthier and more environmentally conscious products and buildings.

The following sections summarize and evaluate some features of the green-building movement related to indoor environmental quality, includ-

[3] The US Federal Trade Commission maintains Guides for the Use of Environmental Marketing Claims (also known as the Green Guides), which are intended to help marketers to make truthful and substantiated claims about "green" products. The guides were under review and revision when the present report was completed at the end of 2010 (FTC, 2010).

ing the process of materials testing and labeling, building-certification programs, and research needed to address knowledge gaps and uncertainties about the effect of the processes as they are related to building design, construction, and operation under future climate scenarios.

There is little peer-reviewed literature on this topic, and the committee's work was informed in part by white papers on building materials and product-testing regimens (Levin, 2010) and green-building rating systems (Srebric, 2010) commissioned by EPA in support of the present study. A 2010 National Research Council workshop report addressed the broader topic of third-party certification systems for products and services labeled as sustainable (NRC, 2010).

Materials Testing and Labeling Systems

There are numerous green-product labeling and whole-building certification systems: a World Resources Institute report indicated that there were more than 340 such systems in November 2009 (WRI, 2010), and a Web site created to compile "ecolabel" information listed nearly 380 in February 2011, including 85 related to building products and 64 to buildings (Ecolabel Index, 2011).

Green-product labeling systems are intended to promote the use of materials that have low or lower problematic emissions as established through uniform laboratory testing at fixed temperatures and airflow rates (Willem and Singer, 2010). Initial protocols for developing voluntary materials-labeling standards through emissions testing were developed in northern Europe—most notably in Germany, Finland, and Denmark—but have since spread around the world. The Levin (2010) and Srebric (2010) white papers and Willem and Singer (2010) and WRI (2010) reviews contain information on the major initiatives; information on some specific programs is highlighted below. International standards and certification programs are addressed because these may affect products used in the United States through imports of certified products, US manufacturers designing for international sales, or competitive pressures for US companies to manufacture green products.

Denmark's DICL[4] system, which was the first to test materials in emission chambers, was developed to address the most common building-related complaints: the evaluation of odors and sensory irritation, particularly irritation of the eyes and upper respiratory tract. It assesses how a standardized exposure to a material or product irritates mucous membranes or is detected by people (Wolkoff and Nielsen, 1995), but it may also incorporate data from animal studies (Wolkoff et al., 1991). Products are labeled in terms of

[4] Danish Indoor Climate Label; also known as the Danish Indeklima Mærke (DIM).

the time that it takes either to achieve no (or a minor) change in breathing frequency in a mouse assay or to drop below odor or irritation thresholds in humans (Kephalopoulos et al., 2005; Wolkoff and Nielson, 1995; Wolkoff et al., 1991). Indoor environment "comfort thresholds" are based on the time required for the VOC emissions to decay to the point where their room concentrations are below their indoor-relevant threshold, which is half the value of either the odor threshold or the sensory-irritation estimate (whichever is lower) for each individual VOC cited in the VOCBASE database (Jensen and Wolkoff, 1996; Kephalopoulos et al., 2005). In most cases, the odor threshold drives the determination of the time value for a specific VOC because sensory-irritation estimates are typically at least an order of magnitude higher than odor thresholds. In practice, use of half the odor threshold is a public-health protective safety factor to account for the presence of the same VOC from other outdoor or other indoor sources in the building. With its focus on irritation and odor thresholds, the DICL test protocol does not address other potential health effects of exposure to hazardous chemicals, such as carcinogenic, allergenic, or endocrine-disrupting properties (Kephalopoulos et al., 2005; Levin, 2010).

REACH (Registration, Evaluation, Authorization, and Restriction of CHemical substances) is the name used for the European Community's (EC's) legislation regarding chemical substances (European Commission Environment, 2011). In contrast to the other programs mentioned here, it has regulatory force with the European Union. The regulation includes provisions regarding emissions from building materials. In 2011, the EC announced that six toxic chemicals, three of which are widely used as plasticizers in flooring, adhesives, and textiles, were being phased out under the authority of REACH (European Commission Environment, 2011).

In 1988, the Canadian government founded EcoLogo—an International Organization for Standards (ISO) Type 1 ecolabel that takes toxicity, recycled content, and renewable energy percentage into consideration but does not consider impacts such as raw material extraction (EcoLogo Program, 2011). It certifies building and construction materials such as adhesives, heating and cooling systems, and paints.

The most well known of the building-materials and furnishing testing systems in the United States may be the certification process promulgated by the GREENGUARD Environmental Institute, an industry-independent nonprofit organization established in 2001 (AQS, 2009b). GREENGUARD's certification processes put building materials and building-related products (such as carpets) into chambers for a fixed period to measure emissions of VOCs, organic acids, formaldehyde, respirable particles, and other compounds. Green Seal is a nonprofit certification organization seeking to reduce the environmental impact of residential and commercial buildings and materials (Green Seal, 2010). It follows guidelines for labeling set by the US government, specifically the Environmental Protection Agency and Federal Trade Commission, awarding a Green Seal to those products that meet its standards. Much of its focus for construction materials is on low VOCs emissions in paints, coatings, and adhesives.

Levin's white paper (2010), which reviews a number of product labeling and certification systems in place in the United States, Europe, and Asia, identifies some weaknesses in their emission testing schemes:

- They measure emissions on a small number of products and test emissions from limited number of samples.
- Results of testing individual products may be used to represent the results of similar products from the same or other manufacturers.
- Tests are performed over a short period that is probably a small fraction of the total service life of many products.
- The focus on building materials, surfaces, and coatings means that other known strong sources, such as consumer products or liquid surface treatments (paints, sealers, and so on), are not systematically tested.

It suggests research and policy initiatives that would help resolve them.

In addition to those issues, a small number of chemicals are measured in the emissions tests, and there are few data on health effects of many of the emitted chemicals in animals or humans with which to assess hazards or develop health-based indoor environmental quality standards (Willem and Singer, 2010). None of the existing or proposed labeling systems includes information on chemical emissions in the text of the labels themselves (Willem and Singer, 2010).

Most important, the testing regimens focus on emissions under "normal" environmental conditions and do not assess the array of potential product-use scenarios, environmental conditions, or air-exchange rates in connection with which building materials or consumer products might be used. For example, the emissions of a mattress in a test chamber held at 50% relative humidity and 72°F (22°C) with one air change per hour will not represent the breathing-zone concentrations experienced by a person sleeping on the mattress where temperatures and humidities are likely to be different. Such an approach also limits the range of interventions that might be undertaken to address problems.

Nonetheless, such testing systems represent an important source of information on product emissions and a driving force in lowering emissions. These will become more important if climate change mitigation and energy conservation measures that encourage tighter buildings become more widespread.

Building Certification

There are numerous green-building or sustainable-building certification systems, including BREEAM (Building Research Establishment's Environmental Assessment Method) in the UK and Canada; CASBEE (Comprehensive Assessment System for Building Environmental Efficiency) in Japan; Green Star in Australia; and Green Globe, GreenPoint, and LEED in the United States (AQS, 2009a; Srebric 2010). Many of them were spawned by green building councils (GBCs), which exist in a number of developed or rapidly developing countries. The World Green Building Council Web site lists 82 nations that have established, associated, emerging, or prospective organizations in early February 2011 (WGBC, 2011). The standards and certifications that the bodies promulgate collectively promote design practices that, in theory, reduce environmental impacts and costs over time, although the evidence base for this assertion is thin (Srebric, 2010).

The US-based GBC's LEED (Leadership in Energy and Environmental Design) certification is a voluntary standard that has become widely accepted for certification of energy efficiency and perceived "greenness" of US buildings (EHHI, 2010; Srebric, 2010). It was one of the first and is among the most widespread standards in the United States. As of September 2010, the organization counted 442 cities, towns, or counties in 45 states; 35 state or territorial governments; and 14 federal agencies or departments that incorporated various LEED initiatives into their regulatory or policy frameworks (LEED, 2010).

The goal of LEED and other rating systems is to provide guidance in the process of building or renovating "green" through a certification or voluntary compliance system. The LEED tiered scoring system for new construction and major renovations awards up to 110 points toward attaining certification at one of four levels: Certified (40–49 points), Silver (50–59), Gold (60–79), and Platinum (80 and above) (LEED, 2011). Seven categories are evaluated:

- Energy and atmosphere (up to 35 points).
- Sustainable sites (26).
- Indoor environmental quality (15).
- Materials and resources (14).
- Water efficiency (10).
- Innovation in design (6).
- Regional "priority" credits (bonus points for water efficiency in the southwestern United States or use of insulation in colder regions, for example) (4).

Those categories are similar to standards employed by GBCs in other countries.

The Srebric white paper (2010) notes that both of LEED's indoor environmental quality prerequisites and 12 of the 15 available points in the category address indoor air quality. The first prerequisite, "minimum indoor air quality performance," is based on compliance with American Society of Heating, Refrigerating and Air-Conditioning Engineers (ASHRAE) building ventilation standards, while the second, "environmental tobacco smoke (ETS) control," is generally achieved by banning smoking in or near the building. Points are also awarded for using low-emitting materials—adhesives and sealants, paints and coatings, flooring systems, and composite wood and agrifiber products—in construction, for indoor chemical and pollutant source control, and for designing for maintaining the thermal comfort of occupants.

From a public-health standpoint, one of the primary criticisms of LEED is that it is possible to receive the highest level of certification without earning any points in indoor environmental quality. Because the system was developed by various stakeholders in the design, materials, and construction industries with little input from the indoor-environment and public-health research communities, point values are weighted more heavily toward the built environment and less toward human exposure and health. The nonprofit organization Environment and Human Health, Inc. (EHHI), published a report in 2010 that offered a number of recommendations for improving the LEED scoring system. They include adding health and environmental-science expertise to the GBC LEED board, simplifying the scoring system and specifying a minimum level of building performance within each of the rating categories, awarding points for the use of known safe products and deducting points for the use of known hazardous substances, and performing postoccupancy indoor air quality testing (EHHI, 2010). The organization also recommended that the GBC take an advocacy role in encouraging federal testing of chemicals used in building products. LEED responded to the EHHI report by acknowledging gaps in its standard but noting that that the criticism discounted the health benefits of buildings' using less energy (Fisher, 2010).

Numerous other organizations also promote standards for various building sectors or in particular regions. The Ecolabel Index lists a number of these (Ecolabel Index, 2011). One such example is *Build It Green*, a California-based nonprofit organization that developed a "GreenPoint" system that rates buildings on resource conservation, indoor air quality, water conservation, community, and energy efficiency (Build It Green, 2011). The system includes consideration of off-gassing of VOCs from building materials and the adequacy of ventilation.

Observations and Synthesis

Product-labeling, emissions-testing, and building-certification systems have the potential to foster the development and use of products and designs that promote environmental stewardship. However, weaknesses in current testing regimens and in the information base on the effectiveness of the systems limit the conclusions that can be drawn about their usefulness in protecting occupant health.

The large number of materials-labeling systems creates confusion because of the lack of standardization, and steps are being taken to address this issue. The European Union is pursuing harmonization among the various European rating systems now in place to ensure consistency within their borders (ECA, 2010). Standardization of protocols for airflow rates and other experimental measures will help foster a milieu where information generated by the testing regimens can more easily be used to predict indoor air concentrations resulting from the use of materials in buildings or consumer products.

Product-labeling systems help to identify products that can contribute to higher scores in green-building certification processes, but the lack of data on types and rates of chemical emissions from materials after installation hampers scientific evaluation of the effect of labeling on indoor environmental quality. That uncertainty also hinders the development of health-based indoor environmental quality standards, as does the lack of research on links to human health and comfort in these buildings.

More specifically, a number of research and information needs are related to emissions testing of materials and consumer products, including development of methods or product-sampling schemes that account for the variability and representativeness of tested building materials; data on source strength of wet-applied products,[5] cleaning products, and air fresheners; and more health-based standards that explore the wide array of potential health effects associated with the products and compounds (Levin, 2010). In addition, because of the current focus on laboratory testing, there have been few studies that validate the utility of testing systems in the real-world environments where materials are used or that recognize that buildings are complex operational units that must simultaneously manage ventilation, moisture, thermal conditions, and other characteristics that affect source strength and exposure. It will thus be necessary to conduct well-designed long-term studies to determine the extent to which emission-testing and building-certification programs and standards achieve improvements in indoor environmental quality.

Buildings are not static: furnishings, equipment, and maintenance prac-

[5] For example, paints, adhesives, sealants, and caulks.

tices change, and structures may be refurbished several times over their operational life. As a result, the contaminants in an extant building may be quite different from the ones that were present when the structure was new. It would be desirable to develop cleaning and maintenance data on the rate of introduction of new materials and furnishings or finishes after occupancy in buildings.

Building ventilation rates vary by season and location, so longitudinal studies that evaluate the variability in emissions over time would be useful. Accurate characterization of indoor chemical concentrations requires numerous samples of a variety of materials, rigorous measurement methods, and accepted quality-assurance and quality-control procedures. Those need to be linked to quantitative work on the effect of LEED and other certification programs on resident health and productivity over time.

Climate change complicates all the problems identified here by introducing more unaccounted-for variables and greater uncertainties. Currently, no building-rating system addresses the effect of changes in future climate conditions even though these changes will certainly affect performance over the lifetime of a structure. To provide clues about performance in a variety of climate change scenarios, future research needs to focus on minimum ventilation rates and room sizes and on scenarios in which to measure emissions. An integrated understanding of the interplay among those factors is crucial for understanding the minimally necessary conditions to maintain healthy indoor environmental quality in a changing climate (Levin, 2010). Reports of respiratory symptoms associated with wallboard from China offer an example of a circumstance in which particular use conditions may have contributed to product breakdown and health problems (Babich et al., 2010; Hooper et al., 2010).

As already noted, material testing and labeling systems represent an important source of information on product emissions and a driving force in lowering emissions. Private sector, federal, and state government efforts are already yielding results as manufacturers seek the advantages that accrue from being able to sell green products. Promoting the use of testing and labeling systems by standards-setting organizations and in the marketplace will accelerate this process, helping to produce healthier indoor environments that are more resilient to the effects of climate change.

HEALTH ISSUES RELATED TO WEATHERIZATION

Energy-efficiency upgrades and weatherization programs have the potential for altering indoor environments of homes. Such measures as sealing ducts, caulking, replacing windows, and increasing insulation may reduce energy consumption, but they may also change airflow patterns, reduce ventilation, and increase moisture and air pollution in a structure.

Effects on Indoor Environmental Quality

Box 8-1 provides examples of potential indoor environmental quality problems resulting from energy-conservation measures in buildings. Additional health or safety issues may arise as new applications are implemented by the home-remodeling industry and the emerging energy-performance industry. Even with the best intentions, indoor environmental quality issues may emerge with interventions that have not been sufficiently well screened for their effects on occupant safety and health.

Researchers have examined the effects of poor ventilation on indoor air quality. Offermann (2010) simultaneously measured indoor and outdoor VOCs, aldehyde, CO_2, and $PM_{2.5}$ levels, and air-exchange rates in 108 newly constructed homes in California. Of the 108 homes, 26 had intermittently operating outside makeup-air systems[6] or continuously operating air-to-air heat exchangers.[7] Some 57% of the homes had 24-h air-exchange rates below the 0.3 ACH recommended in ASHRAE Standard 62.2[8] for residential buildings, and 25% had below 0.18 ACH (Offermann, 2010). The California Office of Environmental Health Hazard Assessment (COEHHA) chronic 8-h reference exposure level (REL) for formaldehyde of 9 $\mu g/m^3$ was exceeded in 98% of the homes (Offermann, 2009). COEHHA's acute-irritation REL of 55 $\mu g/m^3$ was exceeded in 28% of the homes (Wolkoff and Nielson, 2010). Of homes with less than 0.3 ACH, 37% exceeded the 55 $\mu g/m^3$ acute-irritation REL for formaldehyde, and 14% of homes with more than 0.3 ACH exceeded that acute-irritation REL (Offermann, 2010). There was a significant inverse relationship ($p > 0.0001$) between air-exchange rate and formaldehyde concentration. ASHRAE Standard 62.2 allows the use of intermittently operating mechanical ventilation systems. In Offermann's study, homes that had ducted outside-air systems operating intermittently when the heating or cooling systems were on could not maintain sufficient outside air to achieve the minimum ventilation recommendations (Offermann, 2010). All the homes that had continuously operating air-to-air exchangers met ASHRAE Standard 62.2 recommendations.

Offermann (2010) concluded that homes with intermittent outside ducted air did not adequately safeguard occupants against poor indoor air quality, because the homes' coupled fresh-air makeup systems were not operated for long enough periods. That suggested that it would be appropriate to

---

[6] A makeup-air system replaces indoor air exhausted through an HVAC system with outdoor air.

[7] Air-to-air exchangers place indoor air being exhausted from a building and outdoor air being drawn into the building in side-by-side chambers to allow the outdoor air to warm or cool to indoor levels through heat exchange.

[8] ANSI/ASHRAE Standard 62.2-2010, *Ventilation and Acceptable Indoor Air Quality in Low-Rise Residential Buildings*.

BOX 8-1
Examples of Potential Indoor Environmental Quality Problems Resulting from Energy-Conservation Measures in Buildings

Air sealing. Steps taken to make buildings more airtight may lower ventilation rates and, in the absence of source control or the introduction of mechanical ventilation, increase both indoor-air contaminant concentrations and indoor-air humidity. Sealing also has the potential to modify internal air pressure and thus create other problems, such as deficiencies in the makeup air for combustion appliances and exhaust fans. Changing the pressure dynamics in a house can cause depressurization of the foundation or slab and lead to intrusion of soil gases and radon.

Increased insulation. Heavily insulated foundation, wall, and roof systems are more vulnerable to water intrusion, air leakage, and water-vapor migration than more traditional assemblies. Adding insulation to foundations, walls, and roof systems that currently have subacute rain seepage or condensation problems can lead to decay, mold growth, or corrosion problems. Adding insulation to the bottom side of some roof decks or to the inside of brick walls in cold and mixed climates may result in moisture problems.

Some insulation materials may contain irritating chemical compounds, such as formaldehyde in UFFI and some fiberglass insulation and hexabromocyclododecanes (HBCD) in polystyrene insulation (Harrad et al., 2010; Roosens et al., 2009). The long-term durability of spray-on polyurethane foams is of concern because their thermal degradation can generate and release hydrogen cyanide, carbon monoxide, amines, and isocyanates (Carter, 2010).

Building codes in high-risk termite areas often prohibit the use of foam-board insulation on the exterior of a foundation because it interferes with the application of soil pesticide treatments. Foam board on either the interior or exterior of a foundation also makes it difficult to inspect for signs of termite invasion, such as mud tubes (Ogg, 2006). If changes in climatic conditions lead to termites' becoming endemic in areas of the country where they were not previously a problem, then structures that have this form of insulation could be more susceptible to infestation.

High-efficiency combustion equipment. Replacing atmospherically vented combustion equipment (such as furnaces, boilers, and water heaters) in single-family and low-rise multifamily residential buildings with at least 90% efficient combustion or electric equipment lowers the ventilation rate in basements and crawlspaces. In some buildings, that may change the indoor moisture balance and result in cold-weather condensation in the building enclosure. The lowered ventilation rate may also result in increased radon exposure.

Appearance of "legacy hazards." Older homes may have materials that, if disturbed during renovations for energy improvements, can cause health hazards for renovation personnel and occupants. Those materials include asbestos in insulation and tiles and polychlorinated biphenyls (PCBs) in caulking. PCB-containing caulking materials—commonly used in the late 1950s though the 1970s—also pose a liability for owners of buildings constructed during that period, including schools and other public structures.

require airtight energy-efficient homes to have mechanical outdoor-air ventilation that reliably meets or exceeds ASHRAE Standard 62.2. Offermann (2010) calculated that the additional cost for fan power and heating, cooling, and dehumidification for a typical 1,764-ft^2 (164-m^2) home would be $100–300/year, depending on climate region and utility rates.

Epidemiologic Research on Effects of Ventilation on Health and Productivity

As Chapter 4 notes, many studies have examined the relationship between indoor air quality and occupant health, and some of these have considered ventilation—or a proxy for it—among the possible influences. A few studies have directly examined the relationship between ventilation and particular health outcomes or productivity. Mendell (1993), Godish and Spengler (1996), and Seppänen et al. (1999) have all published reviews of the literature. Most recently, Sundell and colleagues (2011) conducted a detailed evaluation of studies of ventilation rate and health or occupant productivity that have appeared in peer-reviewed journals. Table 8-2 summarizes the results of the papers that they reviewed.

The researchers concluded that an association between ventilation rates and health outcomes is biologically plausible. They found that the literature supported links between low ventilation rates and increased risk of allergies, symptoms of sick-building syndrome, and respiratory infections and suggested that "higher rates than are currently common may be health-protective in many instances" (Sundell et al., 2011). However, they noted their conclusions were based on limited data derived primary from colder climates and that there was a great need to collect information from buildings in hot and humid environments.

Several potential explanations of those health outcomes have been put forward. One possibility is that conventional HVAC-system design in air-conditioned buildings—which involves frequently wet surfaces on cooling coils, drain pans, and sometimes humidifiers—may lead to as yet uncharacterized microbiologic exposures and consequent illness (Mendell et al., 2008; Menzies et al., 2003). Poor system condition and poor maintenance increase the risk of such problems. Accumulated dust and dirt and moisture in HVAC systems provide a nutrient source and growth medium for microorganisms (Morey et al., 2009; West and Hanson, 1989).

Ventilation-system hygiene is thus a factor in ensuring good indoor air quality. A 2003 study investigated the health effects of biologic contamination of HVAC systems by examining the association between ultraviolet germicidal irradiation (UVGI) of drip pans and cooling coils in building ventilation systems and indoor microbial concentrations and self-reported symptoms in occupants (Menzies et al., 2003). The researchers system-

TABLE 8-2 Health and Productivity Outcomes Associated with Low Ventilation Rates in Buildings (Adapted from Sundell et al., 2011)

Homes	
Increased allergy symptoms	Bornehag et al., 2005
Increased asthma symptoms	Emenius et al., 2004; Norbäck et al., 1995
Increased bronchial obstruction	Øie et al., 1999
High-occupancy buildings	
Higher rates of respiratory illnesses[a]	Brundage et al., 1988; Hoge et al., 1994; Menzies et al., 2000
Schools	
Degraded perceptions of indoor air quality	Wargocki et al., 2000
Increased symptoms of sick-building syndrome[b]	Wargocki et al., 2002
Increased absences	Milton et al., 2000; Shendell et al., 2004
Decreased performance in school work	Wargocki and Wyon, 2007a,b
Possible reduction in test scores	Shaughnessey et al., 2006
Increased allergy symptoms	Harving et al., 1993; Norbäck et al., 1995; Smedje and Norbäck, 2000; Sundell et al., 1995
Increased asthma symptoms	Smedje and Norbäck, 2000
Increased nasal symptoms	Wålinder et al., 1997a,b, 1998
Office buildings	
Degraded perceptions of indoor air quality	Wargocki et al., 2000
Increased symptoms of sick-building Syndrome[c]	Apte et al., 2000; Erdmann and Apte, 2004; Jaakkola and Miettinen, 1995; Mendell et al., 2005; Stenberg et al., 1994; Sundell et al., 1994a,b; Wargocki et al., 2002
Increased absences	Milton et al., 2000; Shendell et al., 2004
Decreased performance and productivity	Wargocki et al., 2002a, 2004
Increased rhinovirus prevalence	Myatt et al., 2004

[a] Evidence supporting higher rates of respiratory illness in high-occupancy buildings may be the result of confounding factors in addition to low ventilation rates.

[b] Sick-building syndrome (SBS) refers to a combination of nonspecific symptoms related to residence or work in a particular building. Core symptoms may include irritation of the eyes, nose, and throat; cough; dry skin; fatigue; headache; lack of concentration; and high frequency of respiratory tract infections (IOM, 2004). There is no generally agreed-on definition of SBS, and differences in the symptom lists used in various studies make it difficult to draw summary conclusions.

atically turned UVGI lamps installed in the HVAC systems of three office buildings on and off over the course of a year and collected environmental and occupant data. Fungi, bacteria, and endotoxin concentrations were measured, and building occupants who were unaware of the operating condition of the UVGI lamps filled out questionnaires on their health. Other environmental data (temperature, humidity, air velocity, HVAC recirculation, and concentrations of CO_2, nitrogen oxides, ozone, formaldehyde, and total VOCs) and occupant data (participants' assessment of thermal, physical, and air quality and demographic, personal, medical, and work characteristics) were also collected. Occupants reported significantly fewer work-related mucosal symptoms (adjusted odds ratio [OR], 0.7; 95% Confidence Interval [CI], 0.6–0.9) and respiratory symptoms (adjusted OR, 0.6; 95% CI, 0.4–0.9) when the UVGI lamps were on. Reports of musculoskeletal symptoms (0.8; 0.6–1.1) and systemic symptoms (headache, fatigue, or difficulty in concentrating) (1.1; 0.9–1.3) were not significantly different. Although median concentrations of viable microorganisms and endotoxins were reduced by 99% (CI, 67–100%) on surfaces exposed to UVGI, there were no significant decreases in airborne concentrations. The results suggested that limiting microbial contamination of HVAC systems might yield health benefits.

Memarzadeh and colleagues (2010), who studied health-care facilities, cautioned that UVGI disinfection of HVAC systems should not be relied on as the sole intervention used to minimize microbial contamination. The authors stated that

> other factors, such as careful design of the built environment, installation and effective operation of the HVAC system, and a high level of attention to traditional cleaning and disinfection, must be assessed before a health care facility can decide to rely solely on UVGI to meet indoor air quality requirements for health care facilities.

Ventilation effectiveness, the ability of a system to provide supply air that reaches the occupants' breathing zone and distributes conditioned air within occupied spaces to dilute and remove air contaminants (Levin, 1996; NRC, 2006), is one of those factors. The 2006 National Research Council report *Green Schools: Attributes for Health and Learning* found that ventilation effectiveness was—in combination with ventilation rate; filter efficiency; the control of temperature, humidity, and excess moisture; and HVAC operations, maintenance, and cleaning practices—a key factor in good indoor air quality. Experiments by Nielsen (2009) determined that air-distribution patterns with high ventilation effectiveness played an important role in minimizing airborne cross-infection in a hospital setting.

Weatherization Workforce Training in Considerations of Indoor Environmental Quality and Health

As awareness of the potential of weatherization programs to engender problems of indoor environmental quality and health has grown, initiatives have been undertaken to train the weatherization-industry workforce to perform high-quality building retrofits that improve energy efficiency while maintaining or enhancing the health and safety of occupants. These are summarized briefly below.

In January 2011, DOE issued revised guidance for WAP participants aimed at ensuring the health and safety of weatherization workers and recipients of weatherization services (DOE, 2011). The nonprofit National Center for Healthy Housing has developed training programs for a variety of stakeholders, including designers, builders, owners, code inspectors, and public-health workers. Instruction covers new and existing single-family and multifamily buildings and includes ventilation, moisture control, dust control, integrated pest management, material emissions, and management of air-pressure relationships (NCHH, 2008). The American Lung Association of the Upper Midwest's Healthy House program produces guidance for builders and maintains a "Preferred Products" program that lists general cleaning products, stains, finishes, and interior paints that meet standards for low end-use emissions of harmful pollutants and irritants (American Lung Association, 2011).

The *Indoor Air Quality Guide: Best Practices for Design, Construction, and Commissioning* was developed by ASHRAE, the American Institute of Architects, Building Owners and Managers Association International, the Sheet Metal and Air Conditioning Contractors' Association, EPA, and the US Green Building Council. The *Guide* "presents best practices for design, construction and commissioning" and "provides information and tools architects and design engineers can use to achieve an IAQ-sensitive building that integrates IAQ into the design and construction process" (ASHRAE, 2009, p. XII). Its objectives for achieving good indoor air quality comprise the following:

- Manage the design and construction process to achieve good indoor air quality.
- Control moisture in building assemblies.
- Limit entry of outdoor contaminants.
- Control moisture and contaminants related to mechanical systems.
- Limit contaminants from indoor sources.
- Capture and exhaust contaminants from building equipment and activities.
- Reduce contaminant concentrations through ventilation, filtration, and air-cleaning.
- Apply more advanced ventilation approaches.

EPA developed voluntary "Healthy Indoor Environment Protocols for Home Energy Upgrades" and released a draft for public comment in late 2010 (EPA, 2010a). It focuses on indoor environmental quality issues and identifies actions intended to "promote improved occupant health through home energy retrofits" (EPA, 2011a). DOE released a companion draft titled "Workforce Guidelines for Home Energy Upgrades," which details work specifications for high-quality retrofits. Together, the documents are intended to help homeowners, energy auditors, and contractors perform home-energy retrofits that remediate or prevent indoor environmental health problems (EPA, 2010a). If made final and widely adopted, the documents have the potential to define future training efforts.

SYNTHESIS

Buildings are complicated to operate and, to date, operating measures and guidance have been based largely on occupant comfort rather than occupant health or productivity (Sundell et al., 2011). ASHRAE Standards 62.1 and 62.2, for example, offer some guidance for ventilating buildings properly, but these are minimum levels derived from a consensus process based in the engineering, building, and equipment-manufacturing industries. Among the limitations of the current approach are the lack of integration with material-emissions data and the lag between health-science research and guideline or standards-setting processes of government, industry, and consensus groups. There is inadequate understanding of the appropriate indoor air quality design standards or the range of susceptibility inherent in the populations that work in buildings or attend schools. Persily (2010) points out that high-performance guidelines for buildings need to do a better job of addressing the issue of moisture control. Although code requirements and other documents address moisture management, many serious moisture problems persist and—as noted elsewhere in this report—climate change may well exacerbate them.

New building materials and equipment arrive on the market every day, and experience suggests that some may bring unforeseen problems of indoor environmental quality with them. Many such problems might be identified and avoided if the current process of designing and constructing buildings took a more integrated approach that included consideration of the links between indoor environmental quality and indoor and outdoor sources, ventilation, occupant comfort, and energy efficiency.

Material testing and labeling systems represent an important source of information on product emissions and a driving force in lowering emissions. Private sector, federal, and state government efforts are already yielding results as manufacturers seek the advantages that accrue from being able to sell green products. Promoting the use of testing and labeling systems by standards-setting organizations and in the marketplace will accelerate this process, helping to produce healthier indoor environments that are more resilient to the effects of climate change.

Research aimed at developing guidance is needed to address these situations. In addition, weatherization programs should incorporate tracking mechanisms to identify problems of indoor environmental quality problems as they arise and solutions as they are developed and implemented.

CONCLUSIONS

On the basis of its review of the papers, reports, and other information presented in this chapter, the committee has reached the following conclusions regarding building ventilation, weatherization, and energy-use issues:

- Research indicates that poor ventilation in homes, offices, and schools is associated with occupant health problems and lower productivity. However, the information base is inadequate, and studies in hot and humid climates are lacking.
- Proper design, operation, hygiene, and maintenance of HVAC systems contribute to lower microbial contamination, decreased disease incidence, and increased occupant productivity.
- Climate change may make ventilation problems more common or more severe in the future by stimulating the implementation of energy-efficiency (weatherization) measures that limit the exchange of indoor air with outdoor air.
- Government and consensus organizations are beginning to recognize the importance of this issue and have established or are establishing voluntary guidelines and codes that account for the links between energy efficiency, indoor environmental quality, ventilation, and occupant health and productivity. Problems will persist until the weatherization workforce is properly trained to recognize and avoid problems of indoor environmental quality, the efficacy of guidelines and codes is validated, and they are widely implemented.
- Introduction of new materials and weatherization techniques may lead to unexpected exposures and health risks. Energy-efficiency programs must therefore take emissions of building materials and products into account and incorporate tracking mechanisms to identify problems of indoor environmental quality as they arise and solutions as they are developed and implemented.

REFERENCES

AQS. 2009a. *Building rating systems (certification programs): A comparison of key programs.* http://www.aerias.org/uploads/2009.12.09_Green_Building_Programs_Comparison_PUBLISHED.pdf (accessed February 17, 2011).

AQS (Air Quality Sciences, Inc.) 2009b. *Primary green product standards and certification programs: A comparison.* http://www.aerias.org/uploads/2009.03.WP.GreenProdCert ProgCompare.pdf (accessed February 17, 2011).

American Lung Association. 2011. *Healthy house program.* http://www.alaw.org/air_quality/healthy_house_programs (accessed January 19, 2011).

Apte MG, Fisk WJ, Daisey JM. 2000. Associations between indoor CO2 concentrations and sick building syndrome symptoms in US office buildings: An analysis of the 1994-1996 BASE study data. *Indoor Air* 10:246-257.

ASHRAE (American Society of Heating, Refrigerating and Air-Conditioning Engineers). 2009. *Indoor air quality guide: The best practices for design, construction and commissioning.* Atlanta, GA: ASHRAE.

Babich M, Danello MA, Hatlelid K, Matheson J, Saltzman L, Thomas T. 2010. *CPSC staff preliminary evaluation of drywall chamber test results.* US Consumer Product Safety Commission. http://citeseerx.ist.psu.edu/viewdoc/download?doi=10.1.1.161.6951&rep=rep1&type=pdf (accessed February 17, 2011).

Bornehag CG, Sundell J, Hägerhed-Engman L, Sigsgaard T. 2005 Association between ventilation rates in 390 Swedish homes and allergic symptoms in children. *Indoor Air* 15:275-280.

Brundage JF, Scott RM, Lednar WM, Smith DQ, Miller RN. 1988. Building-associated risk of febrile acute respiratory diseases in Army trainees. *Journal of the American Medical Association* 259(14):2108-2112.

Build It Green. 2011. *GreenPoint rated: Your assurance of a better place to live.* http://www.builditgreen.org/greenpoint-rated/ (accessed May 4, 2011).

Carter J. 2010. Making green jobs safe jobs: A case study—The safe use of spray polyurethane foam (SPF). PowerPoint presented at the Good Jobs Green Jobs Conference, Washington, DC. http://www.osha.gov/dep/greenjobs/osha_greenjobs_conference_may2010.ppt (accessed March 4, 2011).

Chan WR, Price PN, Sohn MD, Gadgil AJ. 2003. *Analysis of US Residential Air Leakage Database.* Berkeley, CA: Lawrence Berkeley National Laboratory.

CPUC (California Public Utilities Commission). 2008. *California long term energy efficiency strategic plan: Achieving maximum energy savings for 2009 and beyond.* San Francisco, CA: CPUC.

Cummings JB, Tooley JJ, Jr. 1989. Infiltration rates and pressure differences in Florida homes caused by closed interior doors when the central air handler is on. In *Proceedings of the 14th Passive Solar Conference.* Denver, Colorado: American Solar Energy Society.

Cummings JB, Withers CR, Moyer N, Fairey P, McKendry B. 1996. *Uncontrolled air flow in non-residential buildings.* Orlando, FL: Florida Solar Energy Center.

DOE (Department of Energy). 2009. *2009 buildings energy data book.* Washington, DC: DOE.

DOE. 2010. *Workforce guidelines for home energy upgrades.* http://www1.eere.energy.gov/wip/pdfs/workforce_guidelines_home_energy_upgrades.pdf (accessed January 11, 2011).

DOE. 2011. *Weatherization program notice 11-6.* Washington, DC: DOE.

ECA (European Collaborative Action, Urban Air, Indoor Environment and Human Exposure). 2010. *Harmonisation framework for indoor material labeling schemes in the EU.* Luxembourg: Office for Official Publications of the European. Report No. 27. http://www.eurofins.com/media/1744366/ECA_report_no_27_final%20draft.pdf (accessed February 18, 2011).

Ecolabel Index. 2011. *Ecolabel index: Who's deciding what's green?* http://www.ecolabelindex.com/ (accessed February 17, 2011).

EcoLogo Program. 2011. *EcoLogo Program: Third party certification of environmentally-preferred products.* http://www.environmentalchoice.com/en/ (accessed May 3, 2011).

Edwards JO, Hamilton RI. 1993. Leaks in copper tubing from cooling coils of a large air-conditioning unit. *ASM International, Handbook of Case Histories in Failure Analysis* 2:204-206.

Effinger J, Friedman H. 2010. Right measures. *ASHRAE Journal* 52(10):84, 86, 88-89.

EHHI (Environment and Human Health) Inc. 2010. *The green building debate: LEED certification. where energy efficiency collides with human health.* North Haven, CT: Environment & Human Health, Inc. http://www.ehhi.org/reports/leed/LEED_report_0510.pdf (accessed February 18, 2011).

Emenius G, Svartengren M, Korsgaard J, Nordvall L, Pershagen G, Wickman M. 2004. Building characteristics, indoor air quality and recurrent wheezing in very young children (BAMSE). *Indoor Air* 14:34-42.

Emmerich HJ, McDowell T, Anis W. 2005. *Investigation of the impact of commercial building envelope airtightness on HVAC energy use.* Gaithersburg, MD: National Institute of Standards and Technology.

Emmerich SJ, Persily AK. 2005. *Airtightness of commercial buildings in the U.S.* Gaithersburg, MD: National Institute of Standards and Technology.

Enterprise Community Partners. 2010. *Enterprise green communities.* http://www.greencommunitiesonline.org/about/documents/green_next_gen_fact_sheet.pdf (accessed January 11, 2011).

EPA (Environmental Protection Agency). 2007. *2006 annual report: ENERGY STAR® and other climate protection partnerships.* Washington, DC: EPA.

EPA. 2010a. *Healthy indoor environment protocols for home energy upgrades.* http://www.epa.gov/iaq/pdfs/epa_retrofit_protocols_draft_110910.pdf (accessed January 11, 2011).

EPA. 2010b. *ENERGY STAR.* http://www.energystar.gov/index.cfm?c=about.ab_history (accessed January 11, 2011).

EPA. 2011a. *Draft protocols for home energy upgrades.* http://www.epa.gov/iaq/homes/retrofits.html (accessed January 30, 2011).

EPA. 2011b. *Should you have the air ducts in your home cleaned?* http://www.epa.gov/iaq/pubs/airduct.html (accessed February 28, 2011).

Erdmann CA, Apte MG. 2004. Mucous membrane and lower respiratory building related symptoms in relation to indoor carbon dioxide concentrations in the 100-building BASE dataset. *Indoor Air* 14(Suppl 8):127-134.

European Commission Environment. 2011. *REACH.* http://ec.europa.eu/environment/chemicals/reach/reach_intro.htm (accessed May 4, 2011).

Fischer D. 2010. Do green building standards minimize health concerns? *Scientific American online,* June 7, 2010. http://www.scientificamerican.com/article.cfm?id=do-green-building-standards-minimize-health-cooncerns [*sic*] (accessed February 18, 2011).

FTC (US Federal Trade Commission). 2010. *Request for public comment on proposed, revised guides for the use of environmental marketing claims.* http://www.ftc.gov/os/fedreg/2010/october/101006greenguidesfrn.pdf (accessed February 19, 2011).

Fuller M, Kunkel C, Zimring M, Hoffman I, Soroye KL, Goldman C. 2010. *Driving demand for home energy improvements.* Berkeley, CA: Lawrence Berkeley National Laboratory (LBNL).

Godish T, Spengler JD. 1996. Relationships between ventilation and indoor air quality: A review. *Indoor Air* 6:135-145.

Goldman CA, Fuller MC, Stuart E. 2010. *Energy efficiency services sector: Workforce size and expectations for growth.* Berkeley, CA: Lawrence Berkeley National Laboratory.

Green Seal. 2010. *The original Green Seal of approval since 1989.* http://www.greenseal.org/Home.aspx (accessed May 3, 2011).

Harrad S, de Wit CA, Abdallah MA, Bergh C, Björklund JA, Covaci A, Darnerud PO, de Boer J, Diamond M, Huber S, Leonards P, Mandalakis M, Ostman C, Haug LS, Thomsen C, Webster TF. 2010. Indoor contamination with hexabromocyclododecanes, polybrominated diphenyl ethers, and perfluoroalkyl compounds: An important exposure pathway for people? *Environmental Science & Technology* 34(9):3221-3231.

Harving H, Korsgaard J, Dahl R. 1993. House-dust mites and associated environmental conditions in Danish homes. *Allergy* 48:106-109.

Hoge CW, Reichler MR, Dominguez EA, Bremer JC, Mastro TD, Hendricks KA, Musher DM, Elliott JA, Facklam RR, Breiman RF. 1994. An epidemic of pneumococcal disease in an overcrowded, inadequately ventilated jail. *New England Journal of Medicine* 331(10):643-648.

Hooper DG, Shane J, Straus DC, Kilburn KH, Bolton V, Sutton JS, Guilford FT. 2010. Isolation of sulfur reducing and oxidizing bacteria found in contaminated drywall. *International Journal of Molecular Sciences* 11(2):647-655.

IOM (Institute of Medicine). 2004. *Damp indoor spaces and health*. Washington, DC: The National Academies Press.

Jaakkola JJK, Miettinen P. 1995. Ventilation rate in office buildings and sick building syndrome. *Occupational Environmental Medicine* 52:709-714.

Jensen B, Wolkoff P. 1996. *VOCBASE*. Denmark: National Institution of Occupational Health.

Kephalopoulos S, Koistinen K, Kotzias D. 2005. *Urban air, indoor environment and human exposure. environment and quality of life report No 24: Harmonisation of indoor material emissions labeling systems in the EU. Inventory of existing schemes*. Italy: European Commission. http://www.inive.org/medias/ECA/ECA_Report24.pdf (accessed February 18, 2011).

LEED (Leadership in Energy and Environmental Design). 2010. *LEED public policies. Updated: 09/24/10*. https://www.usgbc.org/ShowFile.aspx?DocumentID=7922 (accessed February 17, 2011).

LEED. 2011. *LEED 2009 for new construction and major renovations. Updated February 2011*. http://www.usgbc.org/ShowFile.aspx?DocumentID=8868 (accessed February 17, 2011).

Levin H. 1996. *Best sustainable indoor air quality practices in commercial buildings*. November 19, 1996, presentation at the Third Annual Green Buildings Conference and Exhibition. San Diego, CA. http://www.buildingecology.com/articles/best-sustainable-indoor-air-quality-practices-in-commercial-buildings/ (accessed March 9, 2011).

Levin H. 2010. *National programs to assess IEQ effects of building material and products*. Washington, DC: EPA Indoor Environments Division. http://www.epa.gov/iaq/pdfs/hal_levin_paper.pdf (accessed February 18, 2011).

Memarzadeh F, Olmsted RN, Bartley JM. 2010. Applications of ultraviolet germicidal irradiation disinfection in health care facilities—Effective adjunct, but not stand-alone technology. *American Journal of Infection Control* 38(5 Suppl 1):S13-S24.

Mendell MJ. 1993. Non-specific symptoms in office workers: A review and summary of the epidemiologic literature. *Indoor Air* 3:227-236.

Mendell MJ, Lei Q, Apte MG, Fisk WJ. 2005. *Outdoor air ventilation and work-related symptoms in U.S. office buildings—Results from the BASE study*. Berkeley, CA: Lawrence Berkeley National Laboratories.

Mendell MJ, Lei-Gomez Q, Mirer AG, Seppänen O, Brunner G. 2008. Risk factors in heating, ventilating, and air-conditioning systems for occupant symptoms in US office buildings: The US EPA BASE study. *Indoor Air* 18(4):301-316.

Menzies D, Fanning A, Yuan L, FitzGerald JM; Canadian Collaborative Group in Nosocomial Transmission of TB. 2000. Hospital ventilation and risk for tuberculous infection in Canadian health care workers. *Annals of Internal Medicine* 133:779-789.

Menzies D, Popa J, Hanley JA, Rand T, Milton DK. 2003. Effect of ultraviolet germicidal lights installed in office ventilation systems on workers' health and wellbeing: Double-blind multiple crossover trial. *Lancet* 362(9398):1785-1791.

Mills E. 2009. *Building commissioning: A golden opportunity for reducing energy costs and greenhouse gas emissions*. Berkeley, CA: Lawrence Berkeley National Laboratory.

Milton DK, Glencross PM, Walters MD. 2000. Risk of sick leave associated with outdoor air supply rate, humidification, and occupant complaints. *Indoor Air* 10(4):212-221.

Morey PR, Rand T, Phoenix T. 2009. On the penetration of mold into the fiberboard used in HVAC ductwork. *Healthy Buildings*, page 4. Syracuse, NY.

Morey PR. 2010. *Climate change and potential effects on microbial air quality in the built environment*. Washington, DC: EPA.

Mowris RJ, Fisk WJ. 1988. Modeling the effects of exhaust ventilation on radon entry rates and indoor radon concentrations. *Health Physics* 54(5):491-501.

Myatt TA, Johnston SL, Zuo Z, Wand M, Kebadze T, Rudnick S, Milton DK. 2004. Detection of airborne rhinovirus and its relation to outdoor air supply in office environments. *American Journal of Respiratory and Critical Care Medicine* 169:1187-1190.

NCHH (National Center for Healthy Housing). 2008. *National Center for Healthy Housing*. http://www.nchh.org/LinkClick.aspx?fileticket=rs1iUR2e%2F%2FA%3D&tabid=298 (accessed January 19, 2011).

Nielsen PV. 2009. Control of airborne infectious diseases in ventilated spaces. *Journal of the Royal Society–Interface* 6(Suppl 6):S747-S755.

NRC (National Research Council). 2006. *Green schools: Attributes for health and learning*. Washington, DC: The National Academies Press.

NRC. 2010. *Certifiably sustainable?: The role of third-party certification systems: Report of a workshop*. Washington DC: The National Academies Press.

Nevius M, Eldridge R, Krouk J. 2010. *The state of the efficiency program industry: Budgets, expenditures, and impacts 2009*. Boston MA: Consortium for Energy Efficiency.

Norbäck D, Björnsson E, Janson C, Widström J, Boman G. 1995. Asthmatic symptoms and volatile organic compounds, formaldehyde, and carbon dioxide in dwellings. *Occupational and Environmental Medicine* 52:388-395.

Offermann FJ. 2009. *Ventilation and indoor air quality in new homes*. Collaborative Report. CEC-500-2009-085. PIER Energy-Related Environmental Research Program. Sacramento, CA: California Air Resources Board and California Energy Commission.

Offermann FJ. 2010. IAQ in airtight homes. *ASHRAE Journal* 52(11):58-60.

Ogg NJ. 2006. *Use of foam insulation below grade may provide access to subterranean termites*. Pendleton, SC: Clemson University.

Øie L, Nafstad P, Botten G, Magnus P, Jaakkola JJK. 1999. Ventilation in homes and bronchial obstruction in young children. *Epidemiology* 10:294-299.

Persily AK, Norford LK. 1987. Simultaneous measurements of infiltration and intake in an office building. *ASHRAE Transaction* 93(2):942-956.

Persily AK, Emmerich SJ. 2009. Effects of air infiltration and ventilation. In *Moisture control in buildings: The key factor in mold prevention—2nd edition*, edited by Trechsel HR, Bomberg MT. West Conshohocken, PA: ASTM International.

Persily AK. 2010. *Indoor air quality guide—Best practices for design, construction and commissioning*. May 14, 2009, presentation to the Federal Interagency Committee on Indoor Air Quality. Washington, DC.

Preziosi P, Czernichow S, Gehanno P, Hercberg S. 2004. Workplace airconditioning and health services attendance among French middle-aged women: A prospective cohort study. *International Journal of Epidemiology* 33:1120-1123.

Rim D, Novoselac A. 2010. Ventilation effectiveness as an indicator of occupant exposure to particles from indoor sources. *Building and Environment* 45(5):1214-1224.

Roosens L, Abdallah MA, Harrad S, Neels H, Covaci A. 2009. Exposure to hexabromocyclododecanes (HBCDs) via dust ingestion, but not diet, correlates with concentration in human serum—Preliminary results. *Environmental Health Perspectives* 117(11):1707-1712.

Seppänen OA, Fisk WJ, Mendell MJ. 1999. Association of ventilation rates and CO_2 concentrations with health and other responses in commercial and institutional buildings. *Indoor Air* 9:226-252.

Shaughnessy RJ, Haverinen-Shaughnessy U, Nevalainen A, Moschandreas D. 2006. A preliminary study on the association between ventilation rates in classrooms and student performance. *Indoor Air* 16(6):465-468.

Shaw CY. 1995. *Maintaining acceptable air quality in office buildings through ventilation*. Ottawa, Ontario: National Research Council Canada Institute for Research in Construction.

Shendell DG, Prill R, Fisk WJ, Apte MG, Blake D, Faulkner, D. 2004. Associations between classroom CO2 concentrations and student attendance in Washington and Idaho. *Indoor Air* 14(5):333-341.

Sherman MH, McWilliams J. 2007. *Air leakage of US homes: Model prediction*. Orlando, FL: Lawrence Berkeley National Laboratory.

Smedje G, Norbäck D. 2000. New ventilation systems at select schools in Sweden—Effects on asthma and exposure. *Archives of Environmental Health* 55:18-25.

Spengler JD. 2010. Personal communication.

Srebric J. 2010. *Opportunities for green building (GB) grating systems to improve indoor air quality credits and to address changing climatic conditions*. Washington, DC: EPA Indoor Environments Division. http://www.epa.gov/iaq/pdfs/jelena_draft_paper_11-4-10.pdf (accessed February 18, 2011).

Stenberg B, Eriksson M, Hoog J, Sundell J, Wall S. 1994. The sick building syndrome (SBS) in office workers. A case-referent study of personal, psychosocial and building-related risk indicators. *International Journal of Epidemiology* 23(6):1190-1197.

Sundell J, Lindvall T, Stenberg B, Wall S. 1994b. Sick building syndrome (SBS) in office workers and facial skin symptoms among VDT-workers in relation to building and room characteristics: Two case-referent studies. *Indoor Air* 4:83-94.

Sundell J, Lindvall T, Stenberg B. 1994a. Associations between type of ventilation and air flow rates in office buildings and the risk of SBS-symptoms among occupants. *Environmental International* 20:239-251.

Sundell J, Wickman M, Pershagen G, Nordvall SL. 1995. Ventilation in homes infested by house-dust mites. *Allergy* 50:106-112.

Sundell J, Levin H, Nazaroff WW, Cain WS, Fisk WJ, Grimsrud DT, Gyntelberg F, Li Y, Persily AK, Pickering AC, Samet JM, Spengler JD, Taylor ST, Weschler CJ. 2011. Ventilation rates and health: Multidisciplinary review of the scientific literature. *Indoor Air* 21(3):191-204.

Wålinder R, Norbäck D, Wieslander G, Smedje G, Erwall C. 1997a. Nasal congestion in relation to low air exchange rate in schools. *Acta Oto-laryngologica* 117:724-727.

Wålinder R, Norbäck D, Wieslander G, Smedje G, Erwall C. 1997b. Nasal mucosal swelling in relation to low air exchange rate in schools. *Indoor Air* 7:198-205.

Wålinder R, Norbäck D, Wieslander G, Smedje G, Erwall C, Venge P. 1998. Nasal patency and biomarkers in nasal lavage—The significance of air exchange rate and type of ventilation in schools. *International Archives of Occupational and Environmental Health* 71:479-486.

WAP (Weatherization Assistance Program). 2009a. *Plans, implementation, and results.* http://www1.eere.energy.gov/wip/plans_implementation_results.html (accessed January 11, 2011).

WAP. 2009b. *Weatherization assistance program.* http://www1.eere.energy.gov/wip/wap.html (accessed January 11, 2011).

WAP. 2010. *The American Recovery and Reinvestment Act and the weatherization assistance program.* http://www.waptac.org/WAP-Basics/Recovery-Act.aspx (accessed January 11, 2011).

Wargocki P, Wyon DP, Ole Fanger P. 2000. Productivity is affected by the air quality in offices. *Proceedings of Healthy Buildings 2000* 1:635-640.

Wargocki P, Lagercrantz L, Witterseh T, Sundell J, Wyon DP, Fanger PO. 2002a. Subjective perceptions, symptom intensity, and performance: A comparison of two independent studies, both changing similarly the pollution load in an office. *Indoor Air* 12(2):74-80.

Wargocki P, Sundell J, Bischof W, Brundrett G, Fanger PO, Gyntelberg F, Hanssen SO, Harrison P, Pickering A, Seppänen O, Wouters P. 2002b. Ventilation and health in non-industrial indoor environments: Report from a European multidisciplinary scientific consensus meeting (EUROVEN). *Indoor Air* 12:113-128.

Wargocki P, Wyon DP, Fanger PO. 2004. The performance and subjective responses of call-center operators with new and used supply air filters at two outdoor air supply rates. *Indoor Air* 14(Suppl 8):7-16.

Wargocki P, Wyon DP. 2007a. The effect of moderately raised classroom temperatures and classroom ventilation rate on the performance of schoolwork by children. *HVAC&R Research* 13(2):193-220.

Wargocki P, Wyon DP. 2007b. The effects of outdoor air supply rate and supply air filter condition in classrooms on the performance of schoolwork by children. *HVAC&R Research* 13(2):165-191.

West M, Hansen E. 1989. Determination of material hygroscopic properties that affect indoor air quality. In *IAQ 89, the human equation: Health and comfort, ASHRAE.* Atlanta, GA: ASHRAE.

WGBC (World Green Building Council). 2011. *Directory of councils.* http://www.mt.worldgbc.org/green-building-councils/gbc-directory (accessed February 17, 2011).

Willem H, Singer BC. 2010. *Chemical emissions of residential materials and products: Review of available information.* Berkeley, CA: Lawrence Berkeley National Laboratory. http://epb.lbl.gov/publications/lbnl-3938E.pdf (accessed February 18, 2011).

Wolkoff P, Nielsen GD, Hansen LF, Albrechtsen O, Johnsen CR, Heinig JH, Franck C, Nielsen PA. 1991. Study of human reactions to emissions from building materials in climate chambers. Part II: VOC measurements, mouse bioassay, and decipol evaluation in the 1–2 mg/m^3 TVOC range. *Indoor Air* 1(4):389-403.

Wolkoff P, Nielsen PA. 1995. A new approach for indoor climate labeling of building materials—Emission testing, modeling, and comfort evaluation. *Atmospheric Environment* 30(15):2679-2689.

Wolkoff P, Nielson GD. 2010. Non-cancer effects of formaldehyde and relevance for setting an indoor air guideline. *Environment International* 36(7):788-799.

WRI (World Resources Institute). 2010. *2010 global ecolabel monitor.* http://www.ecolabelindex.com/downloads/Global_Ecolabel_Monitor2010.pdf (accessed February 17, 2011).

9

Key Findings, Guiding Principles, and Priority Issues for Action

This chapter builds on the foundation laid in Chapters 1–8 to draw out the overarching themes of the report and present its primary recommendations.

OVERVIEW OF THE COMMITTEE'S WORK

The committee's statement of task charged it to summarize the current state of scientific understanding of the effects of climate change[1] on indoor air and public health. The US Environmental Protection Agency (EPA), the report's sponsor, provided three examples of key questions to address:

- What are the likely impacts of climate change in the United States on human exposure to chemical and biological contaminants inside buildings, and what are the likely public health consequences?
- What are the likely impacts of climate change on moisture and dampness conditions in buildings, and what are the likely public health consequences?
- What are the priority issues for action?

While there is substantial scientific literature on the effects of outdoor environmental conditions on the indoors, of indoor environmental condi-

[1] This report uses the term *climate* to refer to prevailing outdoor environmental conditions—temperature, humidity, wind, precipitation, sea level, and other phenomena—and *climate change* to refer to modifications in those outdoor conditions that occur over an extended period of time.

tions on health, of climate change on health, of climate change on buildings, and of buildings on climate change, there is almost no literature on the intersection of climate change, indoor environmental quality (IEQ), and occupant health—and much of what little literature there is summarizes information on one or more of the above categories rather than offering original contributions. The committee was thus required to approach its task by reviewing the available information on components of the climate-change–IEQ–occupant-health nexus and deriving its findings, conclusions, and recommendations and identifying research needs on the basis of a synthesis of that information. It considered peer-reviewed papers, government and research organization reports, and authoritative literature reviews, notably publications in the National Academies' *America's Climate Choices* series (NRC, 2010a,b,c,d), the National Research Council reports *Green Schools: Attributes for Health and Learning* (2006) and *Global Climate Change and Extreme Weather Events: Understanding the Contributions to Infectious Disease Emergence* (2008), and the Institute of Medicine study *Damp Indoor Spaces and Health* (IOM, 2004).

The committee's observations and recommendations are based on general conclusions reached in previous National Academies reports on climate change and literature those reports found to be authoritative. They do not depend on any particular model of future climatic conditions. The literature on IEQ and health is rich and unequivocal: indoor environmental conditions have a great influence on human health, and adverse conditions harm occupant well-being. Altered climatic conditions will not necessarily introduce new risks for building occupants but may make existing indoor environmental problems more widespread and more severe and thus increase the urgency with which prevention and interventions must be pursued.

The committee structured the results of its work into three categories. The **key findings** explicate why people and governments should be concerned about the effects of climate change on the indoor environment. **Guiding principles** are the elements of the public-health mission that informed the specific recommendations offered. The **priority issues for action and recommendations** are the primary initiatives that the committee believes should be implemented to address the problems that it identified. The details underlying these are contained in the preceding chapters.

KEY FINDINGS

Three key findings derived from the committee's literature review underlie its conclusion that alterations in indoor environmental quality induced by climate change are an important public-health problem that deserves attention and action.

Poor indoor environmental quality is creating health problems today and impairs the ability of occupants to work and learn.

There is an extensive scientific literature on the effects of poor indoor air quality, damp conditions, and excessively high or low temperature on human health. Epidemiologic literature reviewed by the committee indicates that pollution intrusion from the outdoors, emissions from building components furnishings, and appliances, and occupant behaviors introduce a number of potentially harmful contaminants into the indoor environment. Dampness problems in buildings are pervasive, and excessive indoor dampness is a determinant of the presence or source strength of several potentially problematic exposures, notably exposures to mold and other microbial agents and to chemical emissions from damaged building materials and furnishings. Damp indoor environments are associated with a number of respiratory and other health problems in homes, schools, and workplaces. Extreme heat has several well-documented adverse health effects. The elderly, those in frail health, the poor, and those who live in cities are more vulnerable to exposure to temperature extremes and to the effects of exposure. Those populations experience excessive temperatures predominantly in indoor environments.

Less information is available on the effects of adverse indoor environmental conditions on the productivity of workers and students. Available studies indicate that inadequate ventilation is responsible for higher absenteeism and lower productivity in offices and schools. Indoor comfort is also important: experiments suggest that work performance and school performance decrease when occupants perceive that a space is too warm or cool or the ventilation rate is too low.

There is inadequate evidence to determine whether an association exists between climate-change–induced alterations in the indoor environment and any specific adverse health outcomes. However, available research indicates that climate change may make existing indoor environmental problems worse and introduce new problems by

- **Altering the frequency or severity of adverse outdoor conditions that affect the indoor environment.**
- **Creating outdoor conditions that are more hospitable to pests, infectious agents, and disease vectors that can penetrate the indoor environment.**
- **Leading to mitigation or adaptation measures and changes in occupant behavior that cause or exacerbate harmful indoor environmental conditions.**

The available research includes

- Models of the potential effects of climate change outdoor conditions and experience with extreme weather events, combined with knowledge of how the outdoor environment influences conditions indoors.
- Measurements of indoor levels of biologic and chemical agents, combined with information on the determinants of high indoor levels and the relationship between outdoor and indoor levels.
- Studies of the association between exposure to biologic and chemical agents or extreme temperature conditions and adverse health outcomes or productivity effects.
- Information and experience concerning the design, construction, operation, and maintenance of buildings and how these affect indoor environmental conditions.
- Studies of the potential health consequences of changes made to buildings as a result of climate change or energy conservation concerns.
- Knowledge of the health consequences of behavioral responses to problems with buildings and their infrastructure.

The lack of directly relevant literature—studies of the intersection of climate change, indoor environmental quality, and occupant health—prevents the committee from drawing more definitive or specific conclusions and underscores the need for the additional data collection and research recommended in this chapter.

Data reviewed as part of the National Academies' *America's Climate Choices* series of reports indicate that global mean temperatures have risen over the past 100 years, heat waves have become longer and more extreme, and cold spells have become shorter and milder. Measurements of rainfall show that moist regions are getting wetter, semiarid regions are becoming drier, and extreme weather events are increasing. Heavier rainfall and earlier thawing and later freezing of rivers and lakes are leading to increased flooding risks. Climate models suggest that those trends will continue and intensify. Such findings are salient for the committee's work because conditions in the outdoor environment help to determine conditions in the indoor environment.

Weather fluctuations and seasonal to annual climate variability influence the incidence of many infectious diseases. Climate change may result in shifting patterns of exposure to pesticides as occupants and building owners respond to infestations of pests like termites whose geographic ranges have changed.

Beginning in the 1970s, rising heating fuel costs created economic pressures to "tighten" buildings to limit heat loss during the winter. Efforts have since expanded to what is now known as the green building movement, which seeks to reduce the energy needed to heat, cool, and light structures and to increase their resiliency in the face of adverse outdoor conditions to limit contributions to and adverse effects of climate change. Such weatherization measures can result in decreased building ventilation rates and—in combination with the introduction of new materials and products indoors—lead to increased pollutant levels indoor and associated adverse exposures in some circumstances.

Other responses to adverse outdoor conditions may also have consequences for indoor exposures and occupant health. Potential increases in the level and frequency of peak electricity demand due to heat waves and in the occurrence of extreme weather events have led to concerns over power outages that could leave building occupants without sources of temperate air and over carbon monoxide poisonings from improper use of generators or other alternative sources of energy and heat.

Opportunities exist to improve public health while mitigating or adapting to alterations in indoor environmental quality induced by climate change.

Although some climate-change adaptation and mitigation measures for the indoor environment have inadvertent adverse health effects, this need not necessarily be the case. Several building technologies, including mixed-mode or hybrid mechanical systems that support natural ventilation, can produce comfortable indoor environments with lower energy costs and greater health benefits than systems typically in use today. Some of them yield additional benefits, such as lower greenhouse-gas emissions or the ability to maintain safe indoor conditions during extended power outages. Widespread introduction of such measures as cool-color building exteriors and appropriate shading, which reduce the amount of heat absorbed by structures, can lower heat-island effects and benefit entire neighborhoods. Such interventions require up-front investments and will vary in their cost-effectiveness depending on the technology, climate, building type and age, and other factors. Inaction also has costs, though, and the public and governments must consider both when deciding whether and how to act.

GUIDING PRINCIPLES

The mission of public health is to "[fulfill] society's interest in assuring conditions in which people can be healthy," and its aim is "to generate organized community effort to address the public interest in health by applying scientific and technical knowledge to prevent disease and promote health" (IOM, 1988). The committee took a public-health approach in formulating its recommendations for reducing the health effects of alterations in IEQ induced by climate change, which can be summarized in three guiding principles:

Prioritize consideration of health effects into research, policy, programs, and regulatory agendas that address climate change and buildings.

Energy-conservation considerations have been the driving force in weatherization-related research. Ventilation guidelines and standards for buildings are based largely on occupant comfort and odor perception. As the country moves toward a future in which climate change will spur the need for increased action to lower buildings' energy demands and increase their resistance to adverse outdoor conditions, it is vital that public health be put in the forefront of the criteria taken into account in making decisions on issues that affect indoor environments.

Make the prevention of adverse exposures a primary goal when designing and implementing climate change adaptation and mitigation strategies.

As *Damp Indoor Spaces and Health* noted, prevention is a foundation principle in public health (IOM, 2004). Indoor environments already present myriad opportunities for exposure to chemical agents in products, outgassing from building materials, emissions from dampness-related microorganisms, airborne pollen and infectious agents, and the like. Common sense suggests that eliminating or lessening those exposures and limiting the introduction of new agents should be the first consideration when responding to potential problems. However, although some steps in that direction have been taken with the introduction of testing regimes to evaluate emissions from building materials and products and the emergence of low-emitting alternatives, more can be done.

Collect data to make better-informed decisions in the future.

A central aim of public-health professionals is "to maximize the influence of accurate data and professional judgment on decision-making—to make decisions as comprehensive and objective as possible" (IOM, 1988). As this chapter has already observed, there is almost no literature at the intersection among climate change, IEQ, and occupant health. It is possible to offer informed views on strategies to minimize the adverse effects of climate change on the basis of existing research, but uncertainties abound, including uncertainties in

- The details of the changes that will take place, the rate at which they are manifested, and their magnitude.
- The influence of technologic advances and other influences on indoor conditions.
- The effects of weatherization and of other adaptation and mitigation measures on public health.

Collecting data that support assessments of the effects of climate change on the indoor environment and health and data on the effects of mitigation and adaptation measures on health will allow future policy to be set in a more informed manner and help to identify misguided or inefficient approaches so that they can be corrected.

PRIORITY ISSUES FOR ACTION AND RECOMMENDATIONS

Chapters 4–8 offer several observations regarding how climate change may affect indoor air quality; dampness, moisture, and flooding; infectious agents and pests; exposure to thermal stress; and building ventilation, weatherization, and energy use. The items below constitute a distillation of the committee's thoughts on how their findings and conclusions should be operationalized.

The committee recommends that the Environmental Protection Agency undertake the following actions.

The Environmental Protection Agency should work with such agencies as the Centers for Disease Control and Prevention to assist state, territorial, and local health and emergency-management agencies in efforts to initiate or expand programs to identify populations at risk for health problems resulting from alterations in indoor environmental quality induced by climate change and to implement measures to prevent or lessen the problems.

EPA is a source of expertise on a number of issues related to the indoor environment and health. The Centers for Disease Control and Prevention (CDC)—which has the lead federal role in monitoring health, detecting and investigating health problems, and developing and implementing responses—already works with EPA on topics of common interest such as the health effects of dampness and mold. Such cooperation will become more important in an era in which extreme weather events are more frequent and severe. EPA's knowledge in such fields as weatherization—where changes in the building envelope may affect ventilation and the presence of moisture indoors and thus IEQ and health—will be of great use in anticipating which future populations may be at risk and in developing solutions. The committee thus recommends that interagency collaboration between EPA and CDC expand into emerging issues of climate change and IEQ. Populations whose health, economic situation, or social circumstances make them more vulnerable to adverse consequences will require special attention in this regard.

The Environmental Protection Agency and other federal agencies should join to develop or refine protocols and testing standards for evaluating emissions from materials, furnishings, and appliances used in buildings and to promote their use by standards-setting organizations and in the marketplace. Standards should include consideration of emissions over the operational life of products and the effects of changes in indoor temperature, dampness, and pests.

Prevention of adverse exposures to materials in the indoor environment and those introduced as a part of weatherization and other climate-change mitigation activities should have high priority, but relatively little information is available. Organizations and government entities in the United States and other countries are pursuing and promoting testing protocols, but the report notes that these efforts are fragmentary. Facilitating the development of uniform test standards not only will let builders and occupants make more informed decisions about which materials, furnishings, and appliances to use in buildings but will simplify compliance for manufacturers.

EPA's Environmental Technology Verification Program and Environmental and Sustainable Technology Evaluations projects, which include a microorganism-resistant building material testing initiative (EPA, 2011a), constitute an example of the agency's current work in this field. Expanded and coordinated action with other federal agencies—including the National Institute of Standards and Technology, which sets testing standards for products and systems and is heavily involved in building research, and the Federal Trade Commission, which is concerned with the accuracy of environmental-product marketing claims—will help to ensure that the resulting protocols are comprehensive and to promote their acceptance.

The Environmental Protection Agency should expand and accelerate its efforts to ensure that indoor environmental quality is protected and enhanced in building-weatherization efforts by facilitating research to identify circumstances in which mitigation and adaptation measures may cause or exacerbate adverse exposures; by reviewing and, where appropriate, changing weatherization guidance to prevent these exposures; and by establishing criteria for the certification of weatherization contractors in health-protective procedures.

One of the primary points made in this report is that buildings are complex systems whose siting, design, and operation interact in ways that are not necessarily easy to predict. Weatherization measures have the potential to inadvertently increase adverse exposures. For example, changes that would reduce ventilation rates would tend to increase indoor radon levels and might also alter the effective radiation dose received.[2] The use of untested building materials could introduce toxic agents to the indoor environment.

EPA and the Department of Energy (DOE) are already cooperating on protocols for home energy-conservation upgrades that were in draft form when the committee completed its report (DOE, 2011b; EPA, 2010). Such recognition of health effects on both occupants and persons performing weatherization work is welcome. It will need to be followed, however, by surveillance activities that evaluate whether guidance is achieving its health-protective objectives and by a mechanism to revise guidance on the basis of evaluation. Certification of weatherization contractors in health-protective procedures would allow consumers to make better-informed decisions on whom they choose to perform work and give governments and utilities guidance on potential service providers.

The research suggested here will take time to yield usable results and, in the interim, EPA will need to use the best available information to inform its judgment on health-protective weatherization policies.

The Environmental Protection Agency in coordination with the Department of Energy, the American Society of Heating, Refrigerating and Air-Conditioning Engineers, and building-code organizations should facilitate the revision and adoption of building codes that are regionally appropriate with respect to climate-change projections and that promote the health and productivity of occupants.

Building codes are predicated in part on local environmental conditions. Codes in northern parts of the country account for the possibility of extended cold and snowy conditions; those in areas prone to hurricanes may require that structures be resistant to extreme weather. If climatic conditions in a particular area change—for example, if there are more severe or more frequent episodes of intense precipitation—buildings constructed under existing codes and designed to operate under previously existing

[2] An investigation conducted by EPA in the 1990s found no consistent relationship between air tightness and indoor radon levels (Dyess, 1994). A large-scale, field study that was under way when this report was completed is revisiting the question, measuring pre- and post-weatherization levels of radon in a nationally representative sample of approximately 550 homes (Tonn et al., 2011).

conditions may fail under the new conditions. That suggests that careful consideration must be given to revising building codes and practices to anticipate future climatic conditions and to taking a coordinated approach to addressing risks.

EPA works in cooperation with the American Society of Heating, Refrigerating and Air-Conditioning Engineers (ASHRAE), a professional organization, in developing guidelines for indoor air quality and ventilation, notably the *Indoor Air Quality Guide: Best Practices for Design, Construction, and Commissioning* (ASHRAE, 2009). ASHRAE standards for building ventilation and thermal comfort are often incorporated in building codes. DOE works with ASHRAE, other professional organizations, industry, and state and local officials on the development and promulgation of building energy codes (DOE, 2011a). ASHRAE, the International Code Council, the US Green Building Council, and the Illuminating Engineering Society of North America joined together to produce an "International Green Construction Code" for potential adoption by regulatory authorities (US Green Building Council, 2010).

The committee recommends that these cooperative efforts on codes be extended to encompass climate-change issues. Most residential and commercial buildings have useful lifetimes that are measured in decades. Promoting research on and development and adoption of regionally appropriate building codes that account for the possibility of future climatic conditions not only will protect the well-being of occupants but could produce economic benefits in the form of longer building lives, lower building insurance fees, and avoided retrofitting costs.

The Environmental Protection Agency and other public agencies and private organizations should join to develop model standards for ventilation in residential buildings and to foster updated standards for commercial buildings and schools. The standards should

- **Be based on health-related criteria.**
- **Account for the effects of weatherization and of other climate-change–related retrofits of existing buildings.**
- **Provide design and operation criteria for mechanical ventilation systems in new construction.**
- **Include consideration of ventilation system hygiene and ventilation effectiveness.**
- **Address how to maintain proper ventilation throughout the life of the system.**
- **Contain "fail-safe" provisions that allow for sufficient air exchange with the outdoors to sustain occupant well-being in the event of ventilation-system breakdown or an extended power outage.**

- **Achieve the objectives mentioned above in an energy- and cost-efficient manner.**

This report has highlighted the central role that ventilation plays in determining IEQ and occupant health. Current ventilation standards, however, are not based on maintaining the health and productivity of occupants and do not account for the potential effects of climate change on building design and operation and on occupant behavior.

The committee believes that action should be taken to address this. There are still information gaps, but the epidemiologic literature makes it clear that poor ventilation in homes, offices, and schools is associated with occupant health problems and lower productivity. Climate change may make ventilation problems more common or more severe by stimulating the implementation of energy-efficiency and weatherization measures that reduce the exchange of indoor air with outdoor air. Because standards are often applied or evaluated only during the initial design process, later changes in the building envelope and the inevitable aging of heating, ventilation, and air-conditioning systems may produce problems in buildings that were initially deemed to have good ventilation. Some states—including California, Connecticut, New York, Minnesota, Vermont, and Washington—already require mechanical ventilation in at least some new construction. That helps to ameliorate ventilation and health concerns but creates a safety risk in circumstances in which failures in building systems or power outages disable mechanical ventilation; this may happen more often if climate change leads to more instances of extreme weather conditions or unsustainable loads on the electric grid due extreme outdoor temperatures.

New ventilation standards should take into account all the considerations listed above. The committee recommends that EPA foster the development and implementation of standards in cooperation with other stakeholders.

The Environmental Protection Agency and other federal agencies should put into place a public-health surveillance system that uses existing environment and health survey instruments to gather information on how outdoor conditions, building characteristics, and indoor environmental conditions are affecting occupant health and on how these change over time.

Chapter 1 lists a number of survey instruments that EPA, DOE, CDC, the Department of Housing and Urban Development, and other government agencies and departments use to gather information on housing characteristics or the health of occupants. Outdoor pollution concentrations, environmental conditions, and climatologic information are separately tracked.

Lack of general population information on the influences of buildings on occupant health hampers the setting of priorities and the development of effective interventions. The committee believes that it is important to start collecting such data. The ideal surveillance system for assessing how climate change affects indoor environment exposures and related health effects would collect data from across the nation and have this clear focus in mind. However, there are substantial logistical hurdles in mounting such an effort, and its high cost may not be tenable under current federal budget circumstances.

The committee therefore recommends that EPA cooperate with its collaborating agencies to identify means for adapting existing environment and health survey instruments to meet the need. All the existing instruments have weaknesses as potential sources of information on the effects of climate change on the indoor environment and health. However, the committee believes that it is possible to identify ways to modify and add to existing instruments such as the National Health and Nutrition Examination Survey (NHANES) and Behavioral Risk Factor Surveillance System (BRFSS) to generate useful data and facilitate combining of databases to perform novel analyses.

The Environmental Protection Agency should exercise a strong level of commitment to educate the public on issues of climate change, the indoor environment, and health. Its efforts should

- **Include materials tailored to those involved in the design, construction, operation, maintenance, and renovation of buildings and to occupants of single-family and multifamily residences.**
- **Consider differences in geography, building type, age, and setting (city, suburb, and rural area) and in current and possible future climate conditions.**
- **Contain specific advice on actions that will reduce the effects of climate change on the indoor environment and will improve health.**

This report began by noting that relatively little attention has been given to the possible effects of climate-change–induced alterations in the indoor environment on occupant health. If adverse effects of climate change are to be prevented, public education and training of professionals will be integral parts of the solution. Education and outreach—especially to those in vulnerable communities and those who provide services to those communities—could have a large role in preventing or limiting adverse effects by making people mindful of potential problems and of the means of addressing them.

EPA already maintains a Web site, *IAQ and Climate Readiness,* that disseminates general information on weatherization, ventilation, and solutions to indoor air-quality problems (EPA, 2011b). The committee recommends that EPA expand its efforts by creating and disseminating specifically tailored messages that speak to the specific circumstances and needs of the diverse audiences listed above and that are focused on steps that these audiences can take to improve IEQ in the spaces that they occupy. It's *Tools for Schools* initiative provides a number of educational products for building professionals, school staff, and the general public aimed at maintaining "a healthy environment in school buildings by identifying, correcting, and preventing [indoor air quality] problems" (EPA, 2011c). These products could be supplemented to cover climate change–related issues. *Tools for Schools* also provides a template for broader outreach on climate change, indoor environment, and health issues for other building types and audiences.

Public health professionals also have a need for education on the issues raised in this report. The public health community is well-versed in how to respond to crises caused by acute circumstances like hurricanes, floods, or heat waves. However, in general, less is known about prevention and control measures for more widespread and chronic issues like building dampness (IOM, 2004). If sanitarians are sensitized to building-related issues and instructed in how to anticipate, identify, and address problematic indoor environmental conditions resulting from climate change, they can add appropriate interventions to their practice and better serve their communities.

Cross-training of those involved in public health and in the design, construction, maintenance, operation, and renovation of buildings in the determinants of good IEQ will help to avoid problems and improve interventions.

The Environmental Protection Agency should continuously evaluate actions taken in response to climate-change–induced alterations in the indoor environment to determine whether they are enhancing occupant health and productivity in a cost-effective manner, should identify initiatives that fail to achieve these objectives, and should take corrective steps as needed.

There is little available research on how changes in climatic conditions may affect the indoor environment. It will therefore be especially important to follow up on the measures taken to lessen adverse effects to determine whether they are effective and whether there are more efficient means of achieving the desired outcomes. The committee therefore recommends that intervention programs include the collection of data that will allow evaluation of whether the programs are materially affecting the health of occupants.

The committee notes that this recommendation is in line with those already offered by the National Research Council's *America's Climate*

Choices: Panel on Informing Effective Decisions and Actions Related to Climate Change. That panel recommended that the federal government "establish information and reporting systems that allow for regular evaluation and assessment of the effectiveness of both government and nongovernmental responses to climate change" and indicated that "decisions and policies should be revised in light of new information, experience, and stakeholder input, and use the best available information and assessment base to underpin the risk management framework" (NRC, 2010c).

The Environmental Protection Agency should spearhead an effort across the federal government to make indoor environment and health issues an integral consideration in climate change research and action plans and to coordinate work on the indoor environment and health.

The serious gap in the scientific literature concerning the relationships among climate change, IEQ, and occupant health identified in this report is a barrier to effective action on the issue. In the committee's judgment, there is a clear lack of recognition of this topic at a level commensurate with its importance.

At the US federal level, the research gap is emblematic of a more fundamental problem regarding indoor environmental health concerns: that responsibility for the integrated environmental, public-health, energy-conservation, housing, urban-planning, and worker well-being issues that make up IEQ do not fall neatly under the aegis of any federal department or agency. Because several organizations have interests in some subjects, yet no entity has the lead responsibility, research needs go unrecognized and unmet, and opportunities for efficient action are unrealized.

The committee believes that this situation must change. Several of the priority issues listed above recommend that EPA either initiate or deepen their cooperation with governmental and other entities on some specific urgent issues, and achievement of their goals will be predicated on building and sustaining robust partnerships. The committee believes that these initiatives should be part of a larger effort to entwine indoor environment and health considerations into the fabric of research and action plans. Because it is difficult to separate the effects of climate change from other influences on the indoor environment, a broad approach to IEQ issues is needed.

There are several potential approaches to addressing the problem.

One is for EPA to initiate action within the US Global Change Research Program (USGCRP)—in which it participates—to address the effects of climate change on indoor environmental quality and on the health and productivity of occupants. The USGCRP, which involves 13 federal departments and agencies, serves as the coordinating body for federal research on climate change and its effects on society (CCHHG, 2011). Major publications of the program do consider the effects of climate change on public health and, separately, on the built environment.[3] However, with few exceptions, public-health considerations are not focused on the indoor environment and health. Discussions of the built environment are centered on threats posed to the infrastructure by flooding and other extreme weather events. The USGCRP is in the process of formulating a new strategic plan with the intent of releasing it in December 2011 (USGCRP, 2011). This process presents an opportunity for EPA to advocate for the inclusion of indoor environment and health concerns into the work of the Program and in particular, the adaptation science; assessments; and communication, education, and engagement elements of the new strategic plan.

EPA should also explore options for stimulating action on climate change, indoor environment, and health issues outside and within the government. These include the initiatives highlighted in the committee's recommendation above that the agency exercise a strong level of commitment to educate the public on these issues.

At the federal level, the committee suggests that EPA promote a broader coordinated effort to address indoor environment and health issues through, for example, the establishment of an interagency working group or a national center. Such mechanisms have been used to effectively coordinate action to identify information gaps, facilitate research, collect data, and catalyze work on other critical issues. An effort to establish a governmental entity to act as a coordinating body will likely require support from the administration or Congress. Nonetheless, the committee believes that consolidating and focusing indoor environmental health efforts may generate efficiencies that make it worthy of consideration and that any efforts that support collaboration in the pursuit of healthy indoor environments will produce societal benefits.

The committee notes that the Public Health Service surgeon general's 2009 *Call to Action to Promote Healthy Homes* already calls for a coordinated federal effort in research, guidance, and technical assistance regarding healthy homes and notes the need for standardization in evaluating interventions (HHS, 2009). The *Call to Action* labels safe and healthy homes as having high federal priority and offers some of the same recommendations put forward in this report, including focusing interventions on the most vulnerable populations and using low-emission building materials.

The United States is in the midst of a large experiment of its own making in which weatherization efforts, energy-efficiency retrofits, and other initiatives that affect the characteristics of interaction between indoor and

[3] In this context, the built environment comprises not only buildings but also the accompanying transportation (roads, bridges, and the like) and public-works (energy, water, sewage, and so on) infrastructures.

outdoor environments are taking place and new building materials and consumer products are being introduced indoors with little consideration of how they might affect the health of occupants. Experience provides a strong basis to expect that some of the effects will be adverse, a few profoundly so. An upfront investment in considering the consequences of these actions before they play out and thereby avoiding problems that can be anticipated would yield benefits in health and in avoiding costs of medical care, remediation, and lost productivity.

REFERENCES

ASHRAE (American Society of Heating, Refrigerating and Air-Conditioning Engineers). 2009. *Indoor air quality guide—Best practices for design, construction, and commissioning*. Atlanta, GA: ASHRAE.

CCHHG (Interagency Crosscutting Group on Climate Change and Human Health). 2011. *Interagency Crosscutting Group on Climate Change and Human Health*. http://www.globalchange.gov/what-we-do/climate-change-health (accessed February 27, 2011).

DOE (US Department of Energy). 2011a. *Building energy codes program*. http://www.energycodes.gov/status/ (accessed February 27, 2011).

DOE. 2011b. *Residential retrofit guidelines*. http://www1.eere.energy.gov/wip/retrofit_guidelines.html (accessed February 27, 2011).

Dyess TM. 1994. *Assessment of the effects of weatherization on residential radon levels*. EPA/600/SR-94/002. Cincinnati, OH: US Environmental Protection Agency, Center for Environmental Research Information.

EPA (US Environmental Protection Agency). 2010. *Healthy indoor environment protocols for home energy upgrades*. http://www.epa.gov/iaq (accessed November 18, 2010).

EPA. 2011a. *Environmental and sustainable technology evaluations (ESTE)*. http://www.epa.gov/etv/este.html (accessed February 27, 2011).

EPA. 2011b. *IAQ and climate readiness*. http://www.epa.gov/iaq/climatereadiness/index.html (accessed February 27, 2011).

EPA. 2011c. *IAQ* Tools for Schools *Program*. http://www.epa.gov/iaq/schools/ (accessed April 26, 2011).

HHS (US Department of Health and Human Services). 2009. *The Surgeon General's call to action to promote healthy homes*. http://www.surgeongeneral.gov/topics/healthyhomes/calltoactiontopromotehealthyhomes.pdf (accessed February 27, 2011).

IOM (Institute of Medicine). 1988. *The future of public health*. Washington, DC: National Academy Press.

IOM. 2004. *Damp indoor spaces and health*. Washington, DC: The National Academies Press.

NRC (National Research Council). 2006. *Green schools: Attributes for health and learning*. Washington, DC: The National Academies Press.

NRC. 2008. *Global climate change and extreme weather events. Understanding the contributions to infectious disease emergence: Workshop summary*. Washington, DC: The National Academies Press.

NRC. 2010a. *Adapting to the impacts of climate change*. Washington, DC: The National Academies Press.

NRC. 2010b. *Advancing the science of climate change*. Washington, DC: The National Academies Press.

NRC. 2010c. *Informing an effective response to climate change*. Washington, DC: The National Academies Press.

NRC. 2010d. *Limiting the magnitude of climate change*. Washington, DC: The National Academies Press.

Tonn B, Rose E, Schmoyer R, Eisenberg JF, Ternes M, Schweitzer M, Hendrick T. 2011. *National evaluation of the Weatherization Assistance Program during the program years 2009-2011*. ORNL/TM-2011/87. Oak Ridge, TN: Oak Ridge National Laboratory.

US Green Building Council. 2010. *ICC, ASHRAE, USGBC and IES announce nation's first set of model codes and standards for green building in the U.S.* Press release dated March 11, 2010.

USGCRP (US Global Change Research Program). 2009. *Global climate change impacts in the United States*. New York: Cambridge University Press.

A

Public Meeting Agendas

PUBLIC MEETING

April 1, 2010
Keck Center of the National Academies
500 Fifth Street, NW, Washington, DC

1:00 p.m.
Conduct of the open session and introduction of participants

John D. Spengler, PhD
Committee Chair

1:05 p.m.
Charge to the Committee

Laura Kolb, MPH
Indoor Environments Division, US Environmental Protection Agency

1:50 p.m.
Open session ends

WORKSHOP 1

June 7, 2010
Keck Center of the National Academies
500 Fifth Street, NW, Washington, DC

11:00 a.m.
Welcome to the National Academies and the Institute of Medicine; conduct of the open session and introduction of participants

John D. Spengler, PhD
Committee Chair

11:15 a.m.
Welcome and opening remarks

Mike Flynn
Director, Office of Radiation and Indoor Air, US Environmental Protection Agency

Session I—Occupant-related issues

11:35 a.m.
Climate change and public health—CDC's perspective and research

Jeremy Hess, MD, MPH
Assistant Professor, Department of Emergency Medicine and Assistant Professor, Department of Environmental and Occupational Health, Emory University Schools of Medicine and Public Health; Consultant, Global Climate Change Program, National Center for Environmental Health, CDC

12:45 p.m.
Vulnerable populations for climate change health effects

John Balbus, MD, MPH
Senior Advisor for Public Health, National Institute of Environmental Health Sciences, CDC

1:10 p.m.
Infectious disease transmission and climate change

David Fisman, MD, MPH, FRCP(C)
Associate Professor, Dalla Lana School of Public Health, and Associate Professor of Health Policy, Management and Evaluation, University of Toronto

1:35 p.m.
The effects of increasing air temperature on humans

Ralph Goldman, PhD
Independent consultant

2:00 p.m.
CO_2, climate change, and the aerobiology of allergenic weeds

Lewis H. Ziska, PhD
Research Plant Physiologist, Crop Systems and Global Change, US Department of Agriculture, Agricultural Research Service

2:25 p.m.
Roundtable discussion—session I speakers

John D. Spengler, PhD, moderator

Session II—Building-related issues

2:50 p.m.
The influence of climatic variables on building and HVAC system design and operation

Andrew K. Persily, PhD
Leader, Indoor Air Quality and Ventilation Group, Building and Fire Research Laboratory, National Institute of Standards and Technology

3:15 p.m.
Adaptation and mitigation strategies for buildings in a changed climate

Terry M. Brennan, MS
President, Camroden Associates, Inc.

3:40 p.m.
Climate change and sustainable architecture

Christoph Reinhart, PhD
Associate Professor of Architectural Technology, Department of Architecture and Group Head, Graduate School of Design–Sustainable Design [G(SD)^2] Initiative, Harvard University

4:10 p.m.
Climate change and the built environment

Franklin W. Nutter, JD
President, Reinsurance Association of America

4:35 p.m.
HUD's national surveys of lead and other residential exposures: A possible model for national IEQ surveillance?

Peter J. Ashley, DrPH
Director, Policy and Standards Division, Office of Healthy Homes and Lead Hazard Control, Department of Housing and Urban Development

5:00 p.m.
Roundtable discussion—session II speakers

John D. Spengler, PhD, moderator

5:15 p.m.
General discussion—day's speakers and committee members

John D. Spengler, PhD, moderator

5:30 p.m.
Workshop ends

WORKSHOP 2

July 14, 2010
Clark Kerr Campus
University of California at Berkeley

9:00 a.m.
Welcome to the Workshop; conduct of the workshop and introduction of the Committee

John D. Spengler, PhD
Committee Chair

9:15 a.m.
Climate change, energy efficiency, and IEQ research

William J. Fisk, MS
Senior Staff Scientist and Department Head, Indoor Environment Department, Lawrence Berkeley National Laboratory

10:00 a.m.
Indoor climate and climate change—a perspective on research needs

Hal Levin, BArch
Research Architect and President, Building Ecology Research Group

11:00 a.m.
The impact of indoor air pollution sources on climate

Kirk R. Smith, PhD, MPH
Professor of Global Environmental Health, and Director of the Global Health and Environment Program, School of Public Health, University of California, Berkeley

11:45 a.m.
Climate change and human health

Kristie L. Ebi, PhD, MPH
Executive Director, Technical Support Unit, Working Group II (Impacts, Adaptation, and Vulnerability), Intergovernmental Panel on Climate Change (IPCC)

1:00 p.m.
Roundtable discussion—committee, speakers, and observers

John D. Spengler, PhD, moderator

2:00 p.m.
Workshop ends

B

Environmental Protection Agency Contractor Reports on Climate-Change, Indoor-Environment, and Health Topics

The US Environmental Protection Agency (EPA) Indoor Environments Division—the sponsor of this study—commissioned a set of white papers on topics related to climate change, the indoor environment, and health to provide information for the committee's consideration. They are listed below[1] and cited, where appropriate, throughout the report. The white papers are also compiled on an EPA Web site that provides links to a number of Agency and contractor reports on issues of indoor air quality (EPA, 2011).

The responsibility for the white papers listed below rests with their authors, and their content does not necessarily represent the views of the committee or the Institute of Medicine.

Contractor Report: Climate Change and Indoor Air Quality

This report presents a general discussion of the effects of climate change on indoor air quality, including occupant influences. Among the issues addressed are how increasing outdoor temperatures may change window and air-conditioning use, moisture intrusion and its adverse health effects, and the effects of weatherization and energy-efficiency efforts on indoor air quality.

Field WR. 2010. *Climate change and indoor air quality.* Washington, DC: US Environmental Protection Agency.

[1] Descriptions of report content are derived in part from EPA (2011).

Contractor Report: Research Needed to Address the Impacts of Climate Change on Indoor Air Quality

This report offers opinions on climate-change and indoor air quality research needs. Topics include high-temperature events; infiltration of outdoor allergens, particulate matter, and ozone; water and dampness intrusion; and disease vectors. The discussion of research gaps focuses on human health but also includes energy efficiency.

Girman J. 2010. *Research needed to address the impacts of climate change on indoor air quality.* Washington, DC: US Environmental Protection Agency.

Contractor Report: National Programs to Assess Indoor Environmental Quality (IEQ) Effects of Building Materials and Products

This report examines national building-materials and product-evaluation programs, which were developed often in response to indoor air quality concerns and vary in focus and scope. These include efforts in the United States, various countries in Europe, the European Union, Japan, and Korea.

Levin H. 2010. *National programs to assess IEQ effects of building material and products.* Washington, DC: US Environmental Protection Agency, Indoor Environments Division.

Contractor Report: Climate Change and Potential Effects on Microbial Air Quality in the Built Environment

This report examines the effects of climate change on pathogens and indoor air quality. Changing climates have caused pathogens and pests to venture into new geographic areas and create new indoor environmental risks, including the possibility of increased pesticide use in response to invading organisms.

Morey PR. 2010. *Climate change and potential effects on microbial air quality in the built environment.* Washington, DC: US Environmental Protection Agency.

Contractor Report: Building Codes and Indoor Air Quality

This report that examines energy-related building codes throughout the United States and how these codes affect ventilation, including air exchange, and indoor air pollution. Ventilation and moisture conditions in existing residential and commercial buildings may be altered because of an increase in extreme weather events due to climate change. Buildings constructed under a set of standards appropriate for the original climate may not be adequate in a different climate.

Mudarri D. 2010. *Building codes and indoor air quality.* Washington, DC: US Environmental Protection Agency.

Contractor Report: Public Health Consequences and Cost of Climate Change Impacts on Indoor Environments

This report addresses the public-health and economic implications of the effects of climate change on indoor environmental quality. It details the effects of biologic agents and of increased humidity, temperature, ventilation, and product emissions on the indoor environment and corresponding human health risks. Climate change and its effects on outdoor contaminants are also examined, and possible adaptation strategies are examined.

Mudarri D. 2010. *Public health consequences and cost of climate change impacts on indoor environments.* Washington, DC: US Environmental Protection Agency.

Contractor Report: Climate Change, Indoor Air Quality and Health

This report describes exposure to common biologic and chemical agents that result from building adaptations. The discussion includes a look at Green Building programs and recommendations on how to make them more considerate of issues of indoor air quality. There is an emphasis on the need for community health-care practitioners to become more involved in addressing susceptible and vulnerable populations.

Schenck P, Ahmed, AK, Bracker A, DeBernardo R. 2010. *Climate change, indoor air quality and health.* Washington, DC: US Environmental Protection Agency.

Contractor Report: Indoor Environmental Quality and Climate Change

This report addresses the impacts of climate change on indoor environments, including material related to potential interventions and solutions.

Brennan T. 2010. *Indoor environmental quality and climate change.* Washington, DC: US Environmental Protection Agency.

Contractor Report: The Impact of Increasing Severe Weather Events on Shelter

This report addresses the impacts of severe weather events on indoor environments. Topics addressed include the use of buildings as shelters from weather extremes.

Brennan T. 2010. *The impact of increasing severe weather events on shelter.* Washington, DC: US Environmental Protection Agency.

DRAFT Contractor Report: Opportunities for Green Building (GB) Rating Systems to Improve Indoor Air Quality Credits and to Address Changing Climatic Conditions

This report describes green-building rating systems, climate change, and indoor environmental quality. Green-building rating systems focus mostly on indoor environments, including moisture, ventilation rates, volatile organic compounds, thermal comfort, and particulate matter but are evaluated in a climate-change context. Two rating systems, those of BREEAM and LEED, are detailed in this report.

Srebric J. 2010. *Draft report: Opportunities for green building (GB) rating systems to improve indoor air quality credits and to address changing climatic conditions.* Washington, DC: US Environmental Protection Agency.

REFERENCE

EPA (Environmental Protection Agency). 2011. *Indoor air—Publications and resources.* http://www.epa.gov/iaq/pubs/ (accessed June 21, 2011).

C

Biographic Sketches of Committee Members and Staff

John D. Spengler, PhD (*Chair*), is the Akira Yamaguchi Professor of Environmental Health and Human Habitation in the Department of Environmental Health of Harvard University's School of Public Health. He has conducted research in personal monitoring, air-pollution health effects, aerosol characterization, and indoor air. More recently, Dr. Spengler has been involved in research that includes the integration of knowledge about indoor and outdoor air pollution and other risk factors into the design of housing, buildings, and communities. He uses the tools of life-cycle analysis, risk assessment, and activity-based costing to measure the sustainable attributes of alternative designs, practices, and community development. Dr. Spengler has served as an adviser to the World Health Organization on indoor air pollution, personal exposure, and air-pollution epidemiology. He serves on the Institute of Medicine Roundtable for Environmental Health and recently chaired a National Research Council Committee on Green Schools. In 2003, Dr. Spengler was the recipient of the Heinz Award for the Environment; in 2008, he was honored by the International Society of Indoor Air Quality and Climate Academy of Fellows with the Max von Pettenkofer award for distinguished contributions to the field of indoor-air science. He received a BS in physics from the University of Notre Dame, an MS in environmental health sciences from Harvard University, and a PhD in atmospheric sciences from the State University of New York-Albany.

John L. Adgate, PhD, is Professor and Chair of the Department of Environmental and Occupational Health in the Colorado School of Public Health, University of Colorado, Denver. His research on exposure assessment, risk analysis, and children's environmental health has focused on improving exposure assessment in epidemiologic studies by documenting the magnitude and variability of human exposures. Dr. Adgate has served on many science advisory panels of the US Environmental Protection Agency, exploring technical and policy issues related to residential exposures. Dr. Adgate received a BA in biology from Calvin College, an MSPH in environmental science from the School of Public Health of the University of North Carolina at Chapel Hill, and a PhD in environmental health granted jointly by the University of Medicine and Dentistry of New Jersey and Rutgers University.

Antonio J. Busalacchi, Jr., PhD, is Director of the Earth System Science Interdisciplinary Center and a Professor in the Department of Atmospheric and Oceanic Science of the University of Maryland. His research interests include tropical ocean circulation and its role in the coupled climate system and climate variability and predictability. Dr. Busalacchi has been involved in the activities of the World Climate Research Programme for many years and is chair of its Joint Scientific Committee. Dr. Busalacchi is chair of the National Research Council Board on Atmospheric Sciences and Climate, a member of its Panel on Advancing the Science of Climate Change, and cochair of the Research Council Committee on National Security Implications of Climate Change on US Naval Forces. He holds a BS in physics and an MS and a PhD in oceanography from Florida State University.

Ginger L. Chew, ScD, is an Epidemiologist in the National Center for Environmental Health of the Centers for Disease Control and Prevention (CDC). She is also Adjunct Professor at Columbia University's Mailman School of Public Health. Dr. Chew's research has focused on exposure assessment of aeroallergens and fungi in the indoor environments of low-income children. She has been part of a team that is designing a nationwide study of low-income homes that have been renovated with green or traditional materials and methods. In 2005, Dr. Chew participated in CDC's environmental-health response to Hurricane Katrina, helping to plan its air-sampling strategy and perform data analysis and interpretation. She holds a BS from the University of Georgia, an MS from the University of Alabama, and an ScD from the Harvard School of Public Health.

Sir Andrew Haines, MBBS, MD, is Professor of Public Health and Primary Care of the London School of Hygiene and Tropical Medicine, where he served as Director until October 2010. His research interests are in epidemiology and health-services research, focusing on the study of environmental influences on health, including the potential effects of global environmental change. In 2009, he chaired an international task force of 55 scientists from nine countries that undertook a program of research on climate-change mitigation and public health, whose results were published in a series of articles in *The Lancet* in December 2009. Dr. Haines serves on a number of major international and national committees, including the Advisory Board of the National Institute for Health Research of England, the Medical Research Council (MRC) Global Health Group, and the MRC Strategy Group. He was formerly a member of the UN Intergovernmental Panel on Climate Change and of the World Health Organization Advisory Committee on Health Research. Dr. Haines earned his MBBS in medicine and MD in medicine and epidemiology from the University of London. He is a Foreign Associate Member of the Institute of Medicine.

Steven M. Holland, MD, is Chief of the Laboratory of Clinical Infectious Diseases of the National Institute of Allergy and Infectious Diseases, National Institutes of Health. He is also a tenured investigator and Chief of the Immunopathogenesis Section of the laboratory. Dr. Holland's major research interests include susceptibility to disseminated and pulmonary mycobacterial infections, mechanisms of mycobacterial and bacterial pathogenesis, and mechanisms of phagocyte immunodeficiency. From 2006 to 2008, he served as President of the International Immunocompromised Host Society. Dr. Holland received his MD from the Johns Hopkins University School of Medicine, where he served as a resident in internal medicine, assistant chief of service in medicine, and fellow in infectious diseases. He is Board-certified in internal medicine with a subspecialty in infectious disease.

Vivian E. Loftness, MArch, FAIA, is University Professor of Architecture at Carnegie Mellon University and Senior Researcher in its Center for Building Performance and Diagnostics. She is an international energy and building-performance consultant for commercial and residential building design and has researched and written extensively on energy conservation, passive solar design, climate, and regionalism in architecture. Prof. Loftness is a member of the Pennsylvania State Climate Change Advisory Committee and has served on several National Academies committees, including the Committee on Review and Assessment of the Health and Productivity Benefits of Green Schools. She has worked for many years with the Architectural and Building Sciences Division of Public Works Canada, researching and developing the issues of total building performance and the field of building diagnostics. Prof. Loftness holds a BS and an MArch from the Massachusetts Institute of Technology. She is a Fellow of the American Institute of Architects and is a registered architect.

Linda A. McCauley, PhD, FAAN, RN, is Professor and Dean of Emory University's Nell Hodgson Woodruff School of Nursing. Dr. McCauley has expertise in the design of epidemiologic investigations of environmental hazards and is nationally recognized for her expertise in occupational-health and environmental-health nursing. Her work aims to identify culturally appropriate interventions to decrease the effects of environmental and occupational health hazards in vulnerable populations, including workers and young children. Dr. McCauley was previously the Associate Dean for Research and the Nightingale Professor of Nursing in the University of Pennsylvania School of Nursing. She received a bachelor of nursing degree from the University of North Carolina, a master's in nursing from Emory, and a doctorate in environmental health and epidemiology from the University of Cincinnati. She is a Member of the Institute of Medicine.

William W. Nazaroff, PhD, is the Daniel Tellep Distinguished Professor and Vice-Chair of the Department of Civil and Environmental Engineering of the University of California, Berkeley. His main research interest is in indoor air quality, with an emphasis on pollutant-surface interactions, transport and mixing phenomena, aerosols, source characterization, exposure assessment, and control techniques; and his teaching activities include a course that assesses the technologic opportunities for mitigating climate change. Dr. Nazaroff is coeditor of *Indoor Air* and Vice President of the International Society of Indoor Air Quality and Climate Academy of Fellows. He received his BA in physics and his MEng in electrical engineering and computer science from the University of California, Berkeley, and his PhD in environmental engineering science from the California Institute of Technology.

Eileen Storey, MD, MPH, is Chief of the Surveillance Branch, Division of Respiratory Disease Studies, National Institute for Occupational Safety and Health, of the Centers for Disease Control and Prevention. She has been serving as Acting Chief for the Surveillance Branch since February 2009. She was formerly Chief of the Division of Public Health and Health Policy and Director of the Center for Indoor Environments and Health at the University of Connecticut Health Center. Dr. Storey's research focuses on the spectrum of respiratory disease associated with indoor environments, with particular interest in the relationship between building-related upper respiratory syndromes, such as rhinitis and sinusitis, and the development of lower respiratory syndromes, such as asthma and hypersensitivity pneumonitis. Her work addresses the development of exposure-assessment tools to characterize indoor risk factors. Dr. Storey received her MD from the Harvard Medical School and her MPH from the Harvard School of Public Health. She is Board-certified in internal medicine and occupational medicine.

INSTITUTE OF MEDICINE STAFF

David A. Butler, PhD, is Senior Program Officer in the Institute of Medicine (IOM). He received his BS and MS in engineering from the University of Rochester and his PhD in public-policy analysis from Carnegie Mellon University. Before joining the IOM, Dr. Butler served as an analyst for the US Congress Office of Technology Assessment, was Research Associate in the Department of Environmental Health of the Harvard School of Public Health, and performed research at Harvard's Kennedy School of Government. He has directed several IOM studies on environmental-health and risk-assessment topics, including ones that produced *Damp Indoor Spaces and Health*, *Clearing the Air—Asthma and Indoor Air Exposures*, *Veterans and Agent Orange: Update 1998* and *Update 2000*, and the series *Characterizing the Exposure of Veterans to Agent Orange and Other Herbicides Used in Vietnam*. Dr. Butler was also a coeditor of *Systems Engineering to Improve Traumatic Brain Injury Care in the Military Health System*.

Lauren N. Savaglio, MS, is a Research Associate in the Institute of Medicine. She received her BS in political science and international relations from Arizona State University and her MS in global health from George Mason University (GMU), where her research interests included pesticide use in agriculture and the nutritional status of those infected with HIV/AIDS. She is also an Adjunct Professor in GMU's Department of Global and Community Health, where she teaches health and environment courses. Before going to the IOM, she practiced as an emergency medical technician at INOVA Fair Oaks Hospital in Virginia, performed HIV/AIDS research for Whitman-Walker Health, and served in the Peace Corps in Togo, West Africa.

Tia S. Carter, MHA, is a Senior Program Assistant in the Institute of Medicine. In December 2008, she graduated with her master's in health-care administration from the University of Maryland, University College. She received her undergraduate degree in community health from the University of Maryland, College Park. Before going to the IOM, she worked as the Health Promotions Coordinator at the Greater Washington Urban League in the Division of Aging and Health Services, where she was responsible for health-promotion and disease-prevention education services and activities among the elderly. She has been involved with the IOM committees responsible for the reports *Asbestos: Selected Cancers* and *Veterans and Agent Orange: Update 2004* and *Update 2006*.

Rachel S. Briks, BS, is a Program Assistant in the Institute of Medicine Board on the Health of Select Populations. She received her BS in community health from the University of Maryland, in College Park in May 2010. Before joining the IOM, she interned at AED Center on AIDS and Community Health and worked as a clerk for the Centers for Disease Control and Prevention National Center for Health Statistics through the Student Temporary Employment Program (STEP).

Victoria Wittig, PhD, was a winter 2010 Christine Mirzayan Science and Technology Policy Fellow at the National Academies. She graduated from the University of Illinois with a PhD in plant biology in May 2008. Her thesis research quantified the effects of two rising greenhouse gases—tropospheric ozone and carbon dioxide—on the growth and productivity of trees, a topic that has implications for understanding and modeling the global carbon cycle and climate change. Previously, Dr. Wittig was a post-doctoral research associate in the Department of Atmospheric Sciences, also at the University of Illinois, where she improved models of photosynthesis to project effects of global changes on the terrestrial carbon cycle. She is now working toward applying her academic training at the intersection of environmental science and public policy in Washington, DC.

Rose Marie Martinez, ScD, is Director of the Institute of Medicine Board on Population Health and Public Health Practice. Before joining the IOM, she was Senior Health Researcher at Mathematica Policy Research, where she studied the effects of health-system change on the public-health infrastructure, access to care for vulnerable populations, managed care, and the health-care workforce. Dr. Martinez is former Assistant Director for Health Financing and Policy with the US General Accounting Office, for which she directed evaluations and policy analysis on national and public-health issues. Dr. Martinez received her doctorate from the Johns Hopkins School of Hygiene and Public Health.